Principles of Ship Design for Polar and Ice Conditions

Although climate change is reducing the extent of sea ice in the polar region, it also enlarges the operation time window and enables more ships to choose routes where ice is prevalent. Global increases in maritime trade add another factor into this. The number of polar ships is on the increase and the total exposure to sea ice is growing significantly. Ships sailing through ice-covered regions need to be powerful enough to overcome the additional resistance, and strong enough to withstand impact with ice. The methodology of designing polar ships, therefore, differs from that for open-water ships.

This book is the first undergraduate and graduate textbook to introduce this topic, ideally for courses on polar ship technology or winter navigation. It sets out the formation process of sea ice, and its physical and mechanical properties, then describes the interaction process between ships and ice for estimating ice load. It introduces methods to evaluate ship performance in ice, including attainable speed and manoeuvrability, then establishes reliable polar ship rules, especially required for ice-strengthening. It sets out experimental methods, both model-scale and full-scale tests, and numerical simulation of ice loading and ship performance.

Pentti Kujala is an emeritus professor of Aalto University Finland and a full tenured professor in Taltech, Estonian Maritime Academy with main research focus on waterway safety. At Aalto University he has served as leader of the Marine and Arctic Technology Research Group, and has also led the Arctic Maritime Operations and Maritime Excellence Research Unit. He has received the European Transport Research Board TRAVISIONS Award for his work to improve the safety of ships in the Baltic Sea and the Arctic.

Fang Li is an associate professor at Shanghai Jiao Tong University, specializing in polar marine engineering, with a particular focus on ice loading on marine structures. He is also an active member of the sea ice committee in Chinese Society for Oceanography.

Mikko Suominen is an assistant professor at Aalto University, Finland, and an expert in ship performance in ice-covered waters and ice-induced loads on ship hulls, with a particular focus on measurement techniques and model-scale testing.

Li Zhou is a professor at Shanghai Jiao Tong University, where he leads the Polar Technology Research Team. He previously worked at Aker Solutions in Norway as an engineer and project manager. He has received the China Shipbuilding Industry Science and Technology Award and the China Institute of Navigation Science and Technology Award, and is an active member of the POAC Committee.

Principles of Ship Design for Polar and Ice Conditions

Pentti Kujala
Fang Li
Mikko Suominen
Li Zhou

Routledge
Taylor & Francis Group
LONDON AND NEW YORK

First published 2027
by Routledge
4 Park Square, Milton Park, Abingdon, Oxon OX14 4RN

and by Routledge
605 Third Avenue, New York, NY 10158

Routledge is an imprint of the Taylor & Francis Group, an informa business

British Library Cataloguing-in-Publication Data
A catalogue record for this book is available from the British Library

ISBN: 978-1-041-14727-5 (hbk)
ISBN: 978-1-041-14921-7 (pbk)
ISBN: 978-1-003-67682-9 (ebk)

DOI: 10.1201/9781003676829

Typeset in Times
by SPi Technologies India Pvt Ltd (Straive)

Contents

About the Authors

Pentti Kujala has more than 45 years of experience in marine safety. He earned his M.Sc. in 1979 and D.Sc. in 1994 from the Helsinki University of Technology. He has worked as a research engineer at Lloyd's Register of Shipping in London (1981–1982), as research scientist at VTT in Finland (1982–1989) and as a project manager at Aker Yards in Finland (2003–2006) and short-term visiting professor at DTU, Memorial University of Newfoundland and TUHH. During 2006–2022, he was a professor of marine technology (safety) at the Aalto University, School of Engineering. He was the head of the Marine Technology research group 2012–2022 and also Vice Dean of Research and Impact at the School of Engineering during 2017–2022. Since 2023, he has been tenured full professor of waterway safety management at Taltech, Estonian Maritime Academy and visiting chair professor at Shanghai Jiao Tong University. He chaired a centre of Excellence for Arctic shipping and operations (CEPOLAR) funded by Lloyd's Register Foundation 2013–2021. He was nominated as the best senior researcher in EU for waterborne research area in 2020. His main research interests have been devoted to the safety and risk analysis of marine operations both in open water and in ice and development of innovative structural solutions for various types of ships. He has published more than 300 publications on these topics.

Fang Li is an associate professor at Shanghai Jiao Tong University. He received his B.Sc. at Dalian University of Technology in China, joint M.Sc. at Norwegian University of Science and Technology in Norway and Aalto University in Finland, and D.Sc. at Aalto University in Finland. His research focused on ice mechanics, polar ship design and navigation safety in polar water. He has been involved in multiple national and international projects related to polar ship design and operation. He has published over 70 peer-reviewed papers related to polar ships.

Mikko Suominen was appointed as an assistant professor in Arctic Ocean Engineering at Aalto University School of Engineering at the beginning of 2024. His research and teaching have focused on ship performance in ice and ice-induced loads on ship hull measurements in full-scale and model-scale since 2008. This has included national and international multidisciplinary research collaboration with academic and private entities during and around expeditions in the Baltic Sea, Arctic, and Antarctic, and measurement campaigns in state-of-the-art mode-scale testing in ice facilities in Finland and Germany. He has been a committee member in the International Towing Tank Conference (ITTC) and International Ships and Offshore Structures Congress (ISSC). He has around 70 peer-reviewed publications in international peer-reviewed journals and conferences.

Li Zhou is a professor at Shanghai Jiao Tong University as team leader of Arctic Technology research group. He has published three books and more than 200 papers in international journals and proceedings. He is now a committee member of the

International Conference on Port and Ocean Engineering under Arctic Conditions (POAC) and International Towing Tank Conference (ITTC). He acted as the chairman of the 11th national ice engineering conference in China in 2022. He earned a Ph.D. in Marine Technology from Norwegian University of Science and Technology and an M.Sc. degree from Shanghai Jiao Tong University. He worked at Aker Solutions as an engineer from 2013 to 2017 with a focus on design of Johan Castberg FPSO, Arctic drillship and platforms in different sizes.

Preface

Shipping activities in ice-covered water and the design of such ships date back to over one century ago. With ongoing climate change, the number of polar ships is on the increase. Ships sailing through ice-covered regions should be powerful enough to overcome the additional resistance due to ice, and meanwhile strong enough to bear the impact with ice. The methodology of designing polar ships, therefore, differs from that for open-water ships. To date, there has not been any textbook to introduce this topic.

This book is intended for those who are interested in the design and operation of polar ships. It can be used as a textbook for undergraduate and graduate courses such as polar ship technology or winter navigation. The readers are expected to understand the fundamental theory, the key aspects and the state of the art for the design of polar ships after reading this book.

The book starts with an introduction of polar shipping and ships. After that, we will introduce the formation process of sea ice, followed by its physical and mechanical properties. Next, in Chapter 3, the interaction process between ships and ice is introduced, which forms the basis of methods for ice-load estimations in Chapter 4. Chapter 5 introduces the methods to evaluate ship performance in ice, including attainable speed and manoeuvrability. Chapter 6 introduces current polar ship rules, especially on ice-strengthening requirements. Chapter 7 introduces experimental methods including both model-scale and full-scale tests. Chapter 8 contains advanced topics on numerical simulation of ice loading and ship performance.

Acknowledgements

The authors would like to thank support from the National Key Research and Development Programme (2024YFC2816300, 2025YFE0125400, and 2025YFE0215000) and National Natural Science Foundation of China (52301331 and 52171259).

1 Introduction to Polar Activities

1.1 POLAR ACTIVITIES

AMSA (2009) gives a summary of the Arctic activities. The Arctic Council, recognizing the critical changes and issues, at the November 2004 Ministerial meeting in Reykjavik, Iceland, called for the Council's Protection of the Arctic Marine Environment (PAME) working group to

> conduct a comprehensive Arctic marine shipping assessment as outlined under the Arctic Marine Strategic Plan (AMSP) under the guidance of Canada, Finland and the United States as lead countries and in collaboration with the Emergency Prevention, Preparedness and Response (EPPR) working group and the Permanent Participants as relevant.

The Arctic Marine Shipping Assessment, or the AMSA 2009 Report, is the product of that Arctic Ministerial decision in Reykjavik and was approved at the 2009 Ministerial meeting in Tromsø.

There is a long history of Arctic marine transport conducted primarily around the ice-free periphery of the Arctic Ocean. Year-round navigation has been maintained since 1978–1979 in the ice-covered western regions of the Northern Sea Route (between the port of Dudinka on the Yenisei River and Murmansk). Previous Arctic marine transport studies for the Northern Sea Route, Canadian Arctic, Alaska's coastal seas and other regions have significant relevance to developing any future regulatory framework for the Arctic Ocean (AMSA, 2009).

The volume of Arctic shipping is on the rise, driven by multiple factors. Arctic holds some of the world's largest remaining oil and gas reserves. In recent years, the extraction of these natural resources has resulted in increased destination-Arctic shipping of cargo in and out of the Arctic. This development is likely to continue even though non-fossil fuel developments has been fast during the recent years, but it will take still at least 10–15 years before we have a good solution outside fossil fuels for maritime transport.

It also expected that the trans-Arctic shipping will increase, especially via the Northern Sea Route (NSR) that offers a significantly shorter distance between Northern Europe and the Far East than the conventional route through the Suez channel. In addition to the expected increase in transport, there is an expected increase in Arctic cruises and tourism. These developments are driven by multiple drivers including climate change, technological development, as well as national and international policies (Bergström et al., 2020).

DOI: 10.1201/9781003676829-1

AIS data analysis gives a good overview of the current shipping activities on the Arctic waters. Figure 1.1 summarize the activities throughout the year. Figure 1.1a gives the summary for the periods: January–March and April–June during 2021 and Figure 1.1b for the periods July–September and October–December. Figure 1.1 clearly indicates the high difference on the traffic volume during the winter and

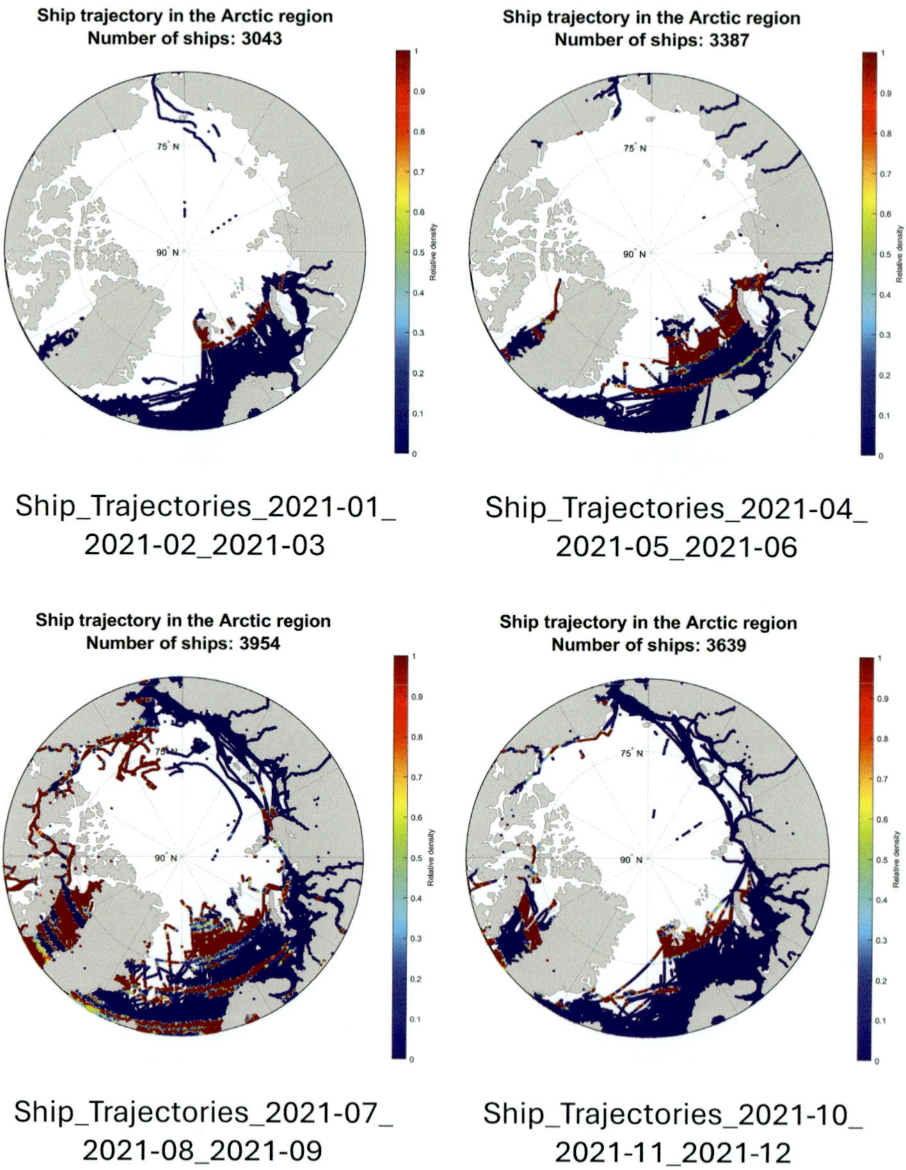

FIGURE 1.1 The traffic intensity on the Arctic waters throughout from 1 Jan to the end of December 2021.

summer period. In winter time, there is in practice only traffic on the northern part of Norway and West part (Until Ob-Bay) of the Russian Arctic. On summer time, there is increasing traffic along the Northern Sea route and also on the Canadian archipelago.

1.1.1 Natural Resources

The Arctic is underlain by sedimentary basins and continental shelves that hold large oil and natural gas resources. Gautier et al. (2009) estimate that the Arctic contains some 13% of the world's undiscovered oil resources and about 30% of its undiscovered conventional natural gas resources. As per Figure 1.2, these resources are distributed between seven major Arctic oil and gas basins: American Basin, Arctic Alaska Basin, East Barents Basin, East Greenland Rift Basin, West Greenland-East Canada Basin, West Siberian Basin and the Yenisey-Khatanga Basin.

In Canadian waters, to date, the development of offshore oil and natural gas production has been limited to sub-Arctic areas offshore Newfoundland, Labrador and Nova Scotia. Currently, no oil and gas drilling is taking place in the Canadian Arctic.

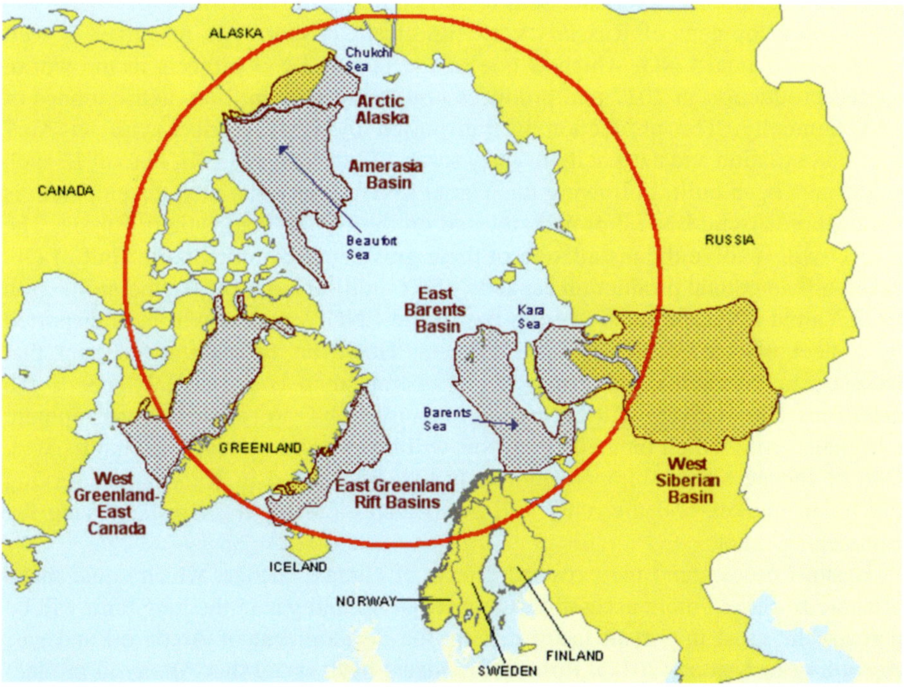

FIGURE 1.2 Major Arctic oil and gas basins.

(Courtesy of U.S. Geological Survey, website: Arctic oil and natural gas resources – U.S. Energy Information Administration (EIA), https://www.eia.gov/todayinenergy/detail.php?id=4650, 13.08.2025.)

In the Norwegian Arctic and sub-Arctic areas, oil and gas exploration and production are presently confined to the Barents Sea. The main developments so far are the Snøhvit and Goliath fields. LNG produced at the Snøhvit field is transported from an onshore terminal at Melkøya outside Hammerfest. For the Goliath field, shuttle tankers are applied. At present, there is one additional field being developed – Johan Castberg – located about 100 km northwest of the Snøhvit-field. Johan Castberg will be developed by a Floating Production Storage and Offloading (FPSO) and Johan Castberg came on stream on 31 March 2025 to produce for 30 years. Potential future fields on the Norwegian shelf, such as the Korpfjell field, are located on areas with seasonal ice. Presently, in Norway, there is a strong political debate about the definition of the so-called ice edge and the strategy for the development of oil and gas fields beyond that edge.

In the Russian Arctic, there is currently a single offshore field – the Prirazlomnoye field – located on the Pechora Sea shelf 60 kilometres off the shore. Oil produced at the field is transported by two 70,000 DWT purpose-built shuttle tankers operated by Sovcomflot. The Prirazlomnoye field still contains over 70 million tonnes of oil and has a potential production level of some 5.5 million tonnes per year. The Prirazlomnoye field will probably not remain Russia's only offshore field for long. As per the Russian Ministry of Natural Resources and Environment, the Russian Arctic shelf holds the equivalent of some 80 billion tonnes of oil, 80% of which is located in the Barents and Kara Seas as shown in Figure 1.2.

The development of Russia's very significant onshore gas resources is well underway. Yamal LNG, which is one of the largest LNG projects in the world, started production in 2017 and produces now some 16.5 million metric tonnes of LNG annually. The production is transported by LNG carriers with an Arc7 ice-class notation and a capacity to carry some 172,000 m^3 of LNG. In total 15 such ships have been built. Following the Yamal LNG project, the next development is the onshore Arctic LNG 2 project, located on the Gydan Peninsula in Russia. The project will involve the installation of three gravity-based units in the Gulf of Ob, each with an annual production capacity of 6.6 million tonnes. Like the production from Yamal LNG, production from the Arctic LNG 2 project will be transported by a fleet of Arc7 ice-class LNG carriers. However, taking into account that ice-class tankers are more expensive to operate than conventional tankers, the use of ice-class tankers will be limited to shuttle LNG to two new transshipment terminals, from which onwards transport will be handled by conventional tankers. One of the new transshipment terminals will be located in Kamchatka, serving the Asian market, and the other will be located close to Murmansk, serving the European market.

Besides politics and the expected effects of climate change, which could make offshore resources more accessible, the price of oil and gas in the long term will be perhaps the most important factor driving the development of Arctic oil and gas. According to Katona (2017), Russian producers have stated that Arctic oil production demands a crude price of at least 70 USD per barrel to be profitable. According to the same source, the Goliath field in Norway is estimated to be profitable at 90 USD per barrel. Anyhow, such estimated breakeven prices appear to be subject to large uncertainty.

Another important activity is mining as there are a number of important minerals on the Arctic. Canada is one active country in this area. Baffinland is waiting on federal approval of a temporary permit that would allow it to ship six million tonnes of iron ore. Baffinland Iron Mines is now looking at 2026 as a start date for its proposed expansion to an iron ore mine in Nunavut.

1.1.2 Destination-Arctic Shipping

The transport of oil and LNG from the Arctic to non-Arctic markets will increase over the next years. In the longer term, the development will depend on the demand and market prices of Arctic natural resources as well as on the costs associated with the exploitation of those. The exploitation costs will depend on national and international policies concerning the exploitation of Arctic natural resources and other natural resources development-related factors.

Given the limitations of the present Oil Spill Response (OSR) in the Arctic, an accident in Arctic waters involving a tanker could have severe consequences both for the environment and the industry. Thus, sustainable development of Arctic natural resources will require a significant strengthening of the available OSR resources, something that will contribute to the overall exploitation costs.

Assuming that the maritime transport costs associated with the exploitation of Arctic natural resources will be relatively low in comparison with the overall exploitation costs, it is probable that the development of destination-Arctic shipping will be rather insensitive to shipping costs.

1.1.3 Trans-Arctic Shipping

There are two major trans-Arctic shipping routes: the NSR and the Northwest Passage (NWP) along the northern coast of North America. As mentioned above, in comparison to the conventional route through the Suez Canal, the NSR provides a significantly shorter route between Northern Europe and the Far East. The NWP, on the other hand, would serve as an alternate route to the Panama Canal, although the distance saved is relatively modest (Matthews et al., 2019). Thus, the consensus is that the NWP will not likely be competitive with existing conventional trading routes (Matthews et al., 2019).

The NSR has been open for international transits since 2009 (Bambulyak et al., 2014). A peak for transit voyages occurred in 2013 with 71 transit voyages (Bambulyak et al., 2014). In 2018, 27 transit voyages were completed along the NSR, including 17 by ships under foreign flags. These figures are very modest in comparison with the estimated potential of the NSR, which has been estimated to be able to carry up to two-thirds of the trade currently passing through the Suez Canal, corresponding to some 8% of the global world trade.

Trans-Arctic shipping is an alternative to conventional shipping. Therefore, the future volumes of trans-Arctic shipping will directly depend on how well trans-Arctic shipping can compete with traditional shipping routes in terms of transport costs and time. We believe that the competitiveness of trans-Arctic shipping is challenging to predict as it depends on several key factors.

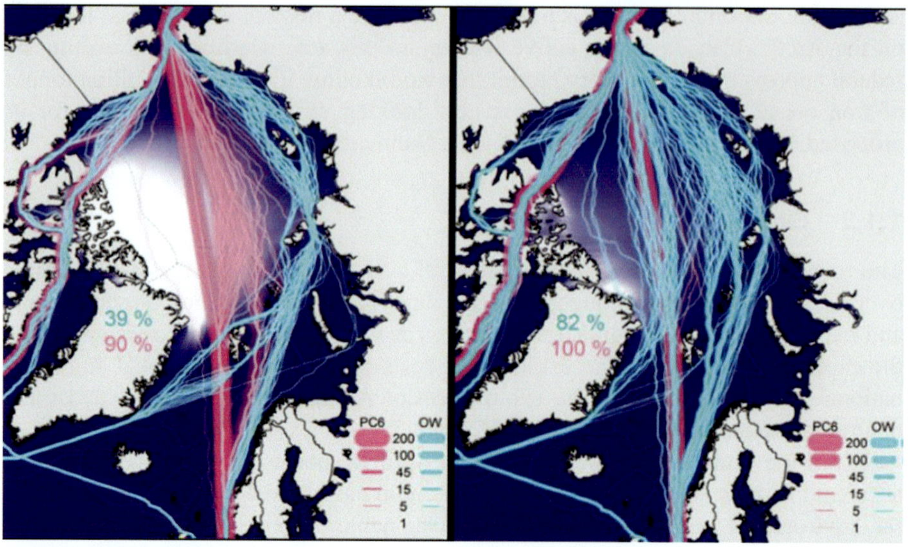

FIGURE 1.3 Fastest available September trans-Arctic routes for future climate scenarios for (a) the period 2015–2030 and (b) 2045–2060. Cyan lines represent open-water (OW) ships, and pink lines represent Polar Class 6 ships capable of navigating in up to 1.2 m thick sea ice. Line weights indicate the number of transits using the same route, and percentages are the probability that Arctic routes are open for the respective ship class. The climate scenarios are calculated for a 'business as usual' scenario where greenhouse gas emissions increase unabated (Melia et al., 2017).

Future environmental change might bring diminishing ice conditions and an extended seasonal ice-free period, affecting future Arctic shipping routes and traffic volumes. Melia et al. (2017) estimate shipping routes and volumes for different future ice scenarios, including those presented in Figure 1.3. As per the figure, diminishing ice conditions might open shorter routes closer to the North Poles. However, full utilization of any ice-free period might be difficult as it is expected to remain challenging to make long enough sea ice forecasts to support the planning of operations.

Regulatory changes limited to Arctic operations, such as a possible HFO ban in Arctic waters could harm the competitiveness of Arctic shipping. On the other hand, regulatory changes also covering non-Arctic shipping, in particular any change resulting in higher energy costs, could help to make Arctic shipping more competitive where it can provide a shorter sailing distance.

Improved navigation and position technology, especially in combination with improved charts and sea ice and weather forecasting, might help masters to select an efficient, safe and economic route. However, improved communication technology might first be needed to support the application of such navigation aids.

As discussed by Bergström (2017), autonomous technologies could be well suited for Arctic cargo ships. Among others, by reducing or eliminating the onboard crew, autonomous technologies could help to reduce safety risks. Autonomous technologies could also help to reduce human errors due to crew fatigue caused by extreme Arctic conditions. Also, by minimizing or removing the onboard crew, related environmental loads such as greywater discharges into the sensitive Arctic waters could be reduced.

National and international policies concerning, for instance, various types of tariffs including icebreaker tariffs, as well as administrative processes, could have a significant effect on the competitiveness of trans-Arctic shipping. Future infrastructure developments concerning, among others, icebreaking and OSR resources are also important. In the longer term, trans-Arctic shipping could even benefit from the construction of new transshipment terminals. Such terminals could limit the use of ice-going ships to distances where sea ice might occur so that the rest of the voyage could be covered by conventional ships that are significantly more cost-effective in open water.

The future price of fuel is another key driver for trans-Arctic shipping. Where trans-Arctic shipping provides a shorter distance to the destination, and hence savings in fuel consumption, it can be assumed that its competitiveness would benefit from a general increase in fuel prices.

1.1.4 Arctic Cruising

The cruise industry has grown by over 60% over the past decade. This is in part because very large non-ice-class ships have started to enter Arctic waters. An example thereof is MV Crystal Serenity, pictured in Figure 1.4, in the summer of 2016 became the largest passenger ship to transit the Northwest Passage. According to Cruise Critic (2020), at least some 25 new polar expedition ships with passenger capacities ranging between around 120 and 530 will enter service before 2022 and this also took place. These ships, many of which will have PC 6 ice-class notation, are expected to contribute to an increase in Arctic shipping in the next few years.

In the medium and longer term, the development of Arctic cruising will largely depend on the associated costs, as well as on its public perception or image. The costs will likely be driven by future regulatory changes and the implementation thereof. For instance, strict enforcement of the Polar Code's five-day post-evacuation survival criteria could be very costly to meet. Likewise, possible future tighter regulations concerning greywater could push costs upwards. On the other hand, because there seems to be a strong market for exclusive luxury-level cruises to cold weather destinations, a high price might limit the market in terms of passenger numbers but not necessarily in terms of value.

Technological developments are also expected to drive Arctic cruising. For instance, improvement in communications technologies could help to increase safety and to improve the passenger experience. Increased safety and efficiency might be provided by improved navigational aids. Improved safety might also be provided by new and improved tools for managing ice loading, which is believed to be critical, especially for cruise ships with a low ice class, and hence presumably a low safety margin with regards to ice loading.

Given the currently limited SAR resources in the Arctic, an accident in the Arctic involving a cruise ship could potentially have catastrophic consequences. Thus, sustainable and safe Arctic shipping requires a significant expansion of the available SAR resources and capabilities in the Arctic.

1.2 POLAR SEA ENVIRONMENT

National Snow and Ice Data Centre (n.d.) has a very good website (https://nsidc.org/arcticseaicenews/) summarizing the development of ice extent on both Arctic and Antarctic waters. Figure 1.4 gives examples of the data available on that website.

Figure 1.4 clearly indicates that the ice extent is decreasing both on the Arctic and Antarctic waters. This means that the shipping activities on these waters will increase as there is less ice and it will be easier to navigate in these waters in future. There is a possibility of an ice-free Arctic Ocean for a short period in summer in future earlier than mid-century (see Figure 1.5). This would mean the disappearance of multi-year ice, as no sea ice would survive the summer melt season. It is highly plausible there will be greater marine access and longer seasons of navigation, except perhaps during winter, but not necessarily less difficult ice conditions for marine operations. There will always be ice, so ships have to be strong and powerful enough to survive in these harsh conditions as they can still encounter heavy ice while navigating on these areas.

FIGURE 1.4 MV crystal serenity courtesy of crystal cruises, 2016.

(Website: crystal-serenity-1024 × 576.jpg (1024 × 576), https://www.cruisetotravel.com/wp-content/uploads/2023/04/crystal-serenity-1024x576.jpg, 13.08.2025.)

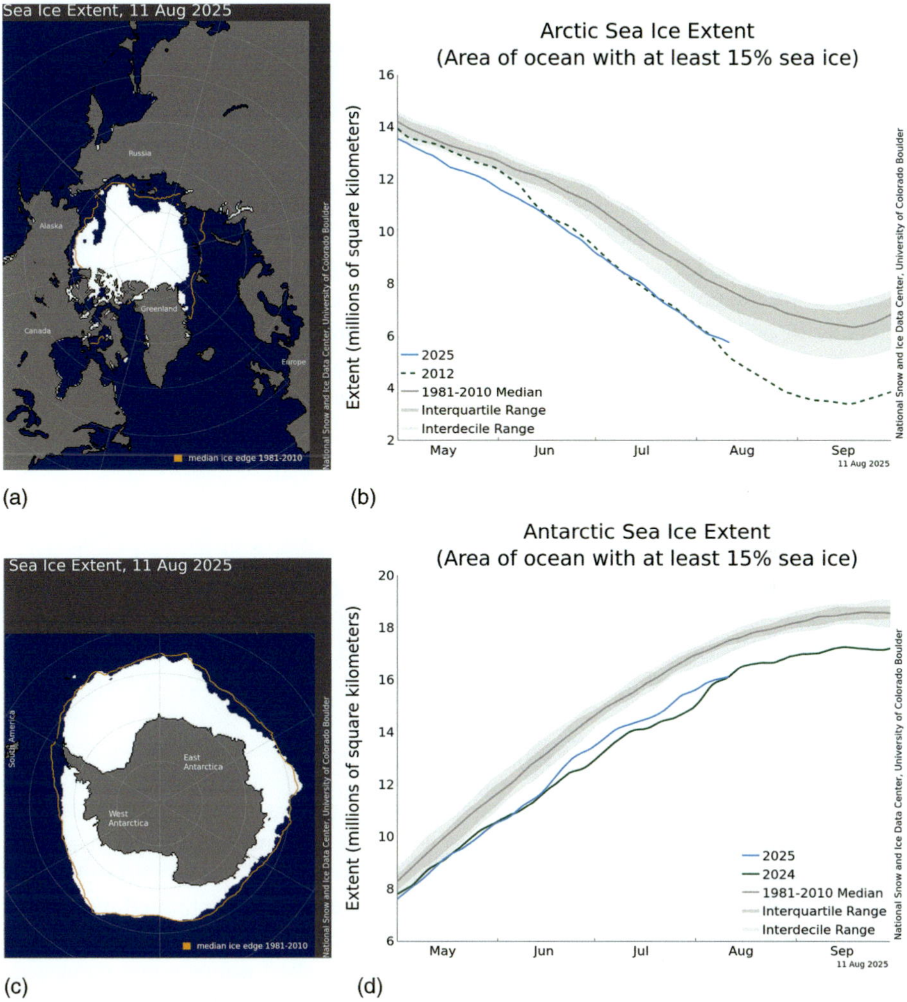

FIGURE 1.5 Examples of ice extent on the Arctic and Antarctic waters.

(https://nsidc.org/arcticseaicenews/, 13.08.2025.)

1.3 POLAR SHIPS

All ship types can be considered important for Arctic shipping: tankers, bulk carriers, offshore supply vessels, passenger ships, tug/barge combinations, fishing vessels, ferries, research vessels and government and commercial icebreakers.

The ship requirements are specified in ship detail description including ship operation requirements and planned operation area. The ice performance requirement is specified in detail description of the planned ship. Strength requirements are typically based on selected ice class of the ship. Ship operation area specification included typically: Minimum temperature on the operating area, planned time in icy waters and cold weather during the operations and autonomous time required for the ship.

Various ship types have different operation requirements and typically the ship types are divided into three categories: icebreakers, ice-going ships and ice-strengthened ships. Icebreakers have to be able to navigate independently on all the planned ice conditions and they should never get stuck in ice. The hull has to be strong enough for all the planned ice conditions. Ice-going ships can navigate independently in ice on the planned operation area. Ice-strengthened ships are competing with no ice-strengthened ships in open water as most of the time they navigate in open water, therefore they will have compromised open water and ice-going performance. They can be capable of breaking thin ice but typically they need icebreaker assistance on normal operation conditions. Figure 1.7 summarizes typical ships on each category.

1.4 SHIP OPERATION PRINCIPLES IN ICE

Individual ship performance in ice is described in Chapter 5 of the text book. Here we present the principles of icebreaker assistance and ice management, which are two typical important operations in ice. For both of these, the system-level understanding of the ice conditions, ice dynamics and ship performance in ice are important as described further next.

1.4.1 Icebreaker Assistance Principles

The icebreaker assistance principles are somewhat different on various sea areas. In Canadian and US waters, icebreakers do not assist systematically cargo vessels as it is done in the Baltic Sea and Russian Arctic. In Antarctica, there is no icebreaker assistance available. The Canadian and US Coast Guards have a limited number of icebreakers available for the escort and support of shipping. These icebreakers are heavily committed and cannot always be provided on short notice when requested. Their main duty is to help only when a ship gets stuck in ice or have another urgent need of help.

Icebreaker assistance is typically required for ice-strengthened ships navigating in the Norther Sea Route along the Russian coast, see Figure. 1.8 as an example of a convoy in ice. The principles of icebreaker assistance in the Russian Arctic have been studied by Topaj and Tarovik (2023). They investigate the problem of using the operative planning instruments in the simulation models of Arctic. It is claimed that the presence of the operational planning algorithm significantly influences the behaviour of a simulated system even at the early design stage. The paper describes the idea of incorporating an external optimal planning engine into the simulation model. The approach is demonstrated with two simplified examples: icebreaker fleet management on one-dimensional graph and planning of oil transshipment using floating storage facilities. The paper also contains the comparison of the proposed planning approach with the alternative simplified solutions, such as the situational algorithm for local choice of proper resource for each arrived agent. The obtained

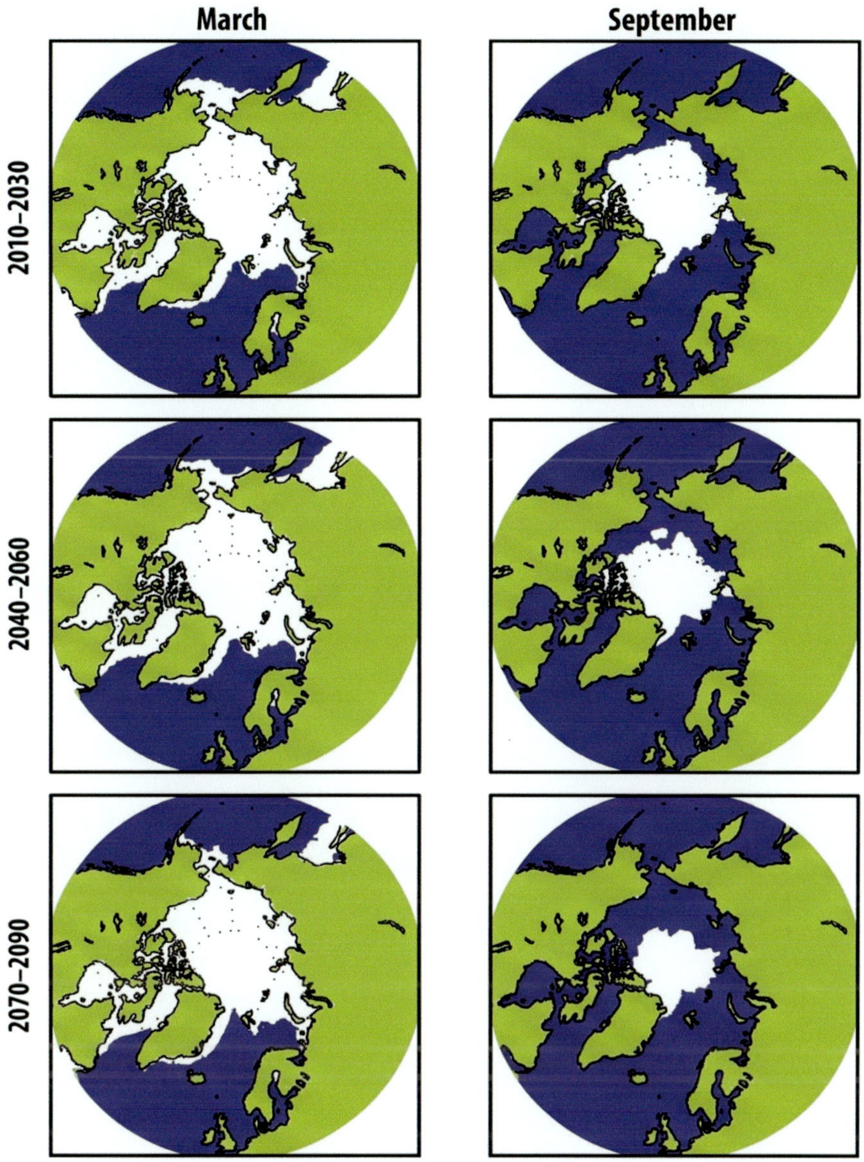

FIGURE 1.6 Arctic Sea ice simulations for the 21st century.

(AMSA, 2009, picture on page 27.)

results made it possible to conclude that the incorporation of rather complex combinatorial optimization algorithms into simulation model leads not only to more adequate but also to a decrease in the stochastic volatility of simulation results.

Finland has more than 150 years' experience of winter navigation, called today as Finnish-Swedish Winter Navigation System (FSWNS). The FSWNS has been

FIGURE 1.7 Typical ships types for navigating in ice: (a) Baltic icebreaker *Polaris*; (b) Polar icebreaker *Fennica* (c) Polar icebreaker *Xuelong II*; (d) Ice-going ship *S.A. Agulhas II*; (d–f) Ice-strengthened ships following an icebreaker or being towed (1.7e).

(a) (b)

FIGURE 1.8 Examples of ship convoys in Russian Arctic. MT *Uikku* is assisted by IB *Kapitan Dranitsyn* and Nuclear IB *Rossiya*.

developed to guarantee safe and reliable shipping throughout the winter even though all of the Finnish harbours are ice-covered each winter. The system consists of three elements: ice-strengthened ships, icebreaker assistance and traffic restrictions. The traffic restrictions ensure that only ships with some minimum size and ice class will get icebreaker assistance on the winter period and the harder the ice conditions the stricter will be the traffic restrictions. This is to guarantee that the icebreaker fleet is capable of offering the service required to keep the winter navigation operations safe and aiming for minimum waiting time for the merchant vessels. In order to ensure ships possess the necessary ice-going capabilities for safe and efficient operations, adherence to the Finnish-Swedish Ice Class Rules (FSICR) is imperative.

Aalto University in a close cooperation of the Finnish Transport and Communication Agency has developed during the last ten years a simulation tool to study the system-level performance of the winter navigations in Finland (Kujala et al., 2024; Kulkarni et al., 2022a, 2022b, 2024). The main aim of the tool development has been to obtain a research-based prototype and decision-making aide that can be used to simulate the need of icebreakers as well as socio-economical and environmental impacts related to FSWNS in the future when both ice conditions, maritime traffic and characteristics of ice-strengthened ships will be under dynamic change. Climate change will affect the ice conditions so that the maximum ice thickness might decrease but the annual variations will be high and the ice movements can increase causing other type of challenges for the winter navigation system. Ship's size seems to increase and at the same time the new strict environment regulations will decrease the used engine power on ships. In efforts to curb Greenhouse Gas emissions from ships, the International Maritime Organization implemented the Energy Efficiency Design Index (EEDI) regulations in 2011. These regulations are geared towards promoting the adoption of highly energy-efficient solutions in maritime operations. While the focus is on reducing GHG emissions, the technical specifications of these regulations restrict the maximum propulsion power installed on ships and encourage

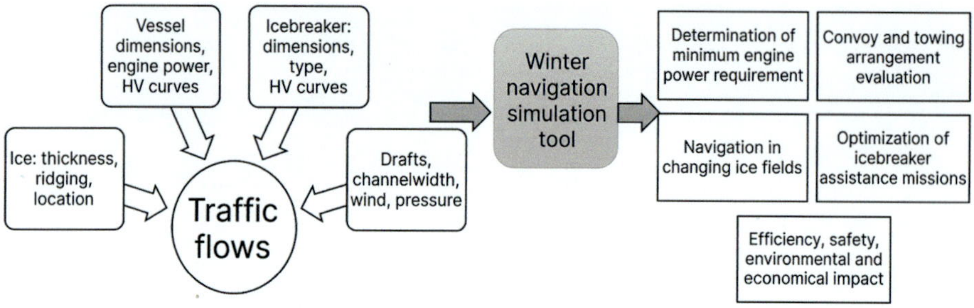

FIGURE 1.9 Simulation tool structure (Kulkarni et al., 2022b).

the use of hull forms optimized for open water conditions. This will decrease dramatically the ice-going capabilities of these ships. All these aspects are included in the simulation tool (see Figure 1.9).

1.4.2 Ice Management Principles

Ice management to support station keeping in ice has been conducted in oil exploration projects since 1976 in the Canadian Beaufort Sea and later in the U.S. Beaufort Sea to manage the ice and extend the drilling season of drill ships primarily intended for open water use (Haimelin et al., 2017). A conical drilling unit, the Kulluk, which was designed to withstand significant ice loads, started drilling operations in the Canadian Beaufort Sea in 1983. The operation was supported by anchor handling and supply icebreakers also performing physical ice management when required. The first ice management operation to support Dynamically Positioned (DP) station keeping was conducted offshore Sakhalin Island in 1999 to support diving and construction operations (Keinonen et al., 2000). The first, and so far only, ice management supported operation, where oil has been produced, was also offshore Sakhalin Island (Keinonen et al., 2006b). A scientific Arctic Coring Expedition, trying the limits of an ice management system and DP station keeping in ice, was conducted in 2004 in the central polar pack (Keinonen et al., 2006a).

Iceberg management is seen as one form of ice management and it concerns mainly operations in open water, but in an area where iceberg incursions can occur. The first iceberg management techniques were developed in offshore eastern Canada oil exploration projects already from the early 1970s. Year-round oil production started in 1997 at the Hibernia field offshore Newfoundland and iceberg management has been an important part of the activity since. Eik (2008) has written a comprehensive review on the past projects involving ice and iceberg management. Most of the past projects involving ice management have been about petroleum activities. This is apparent as very few offshore or marine activities have such potential gains that would justify the use of costly supporting operations such as ice management. Hamilton et al. (2011) calculate that approximately half of the estimated 82 billion barrels of Arctic oil will lie deeper than 100 m water depth, requiring floating exploration and production systems and consequently a supporting ice management operation. Kubat and Sayed (2014) have written a wide literature review concentrating on the technological challenge of station keeping in ice.

Studying the risks related to an operation in a structured manner requires system thinking. The usually many components and complex processes involved in an operation can be classified and systematized in several ways, depending on the viewpoint and purpose for doing so. Here, the viewpoint is a bird's eye view, looking at the ice management system objectively from above. As each ice management operation is unique and there is no such thing as a standard operation, the systematization here is kept at a high level to be applicable to different operations.

The system elements are identified by Haimelin et al. (2017) from literature and by interviewing people who have operational ice management experience. They are categorized based on their type or nature, so that specific methods could be applied to identify and quantify the risks related to a specific category. Although the ice management system is considered here as a separate system or a sub-system of the main activity, in reality it works as an integrated part of the main activity or at least in very close co-operation with it. The separation made here is a logical boundary between the ice management operation and the main activity. The ice management system defined here has inputs, which it processes through system functions and which further exit the system as outputs. Figure 1.10 illustrates the ice management system. The system inputs are divided into two categories: the ice input and the information input. The ice input consists of the parameters of the sea ice that is to be managed. These are the encountered ice properties, such as ice thickness, coverage, strength, floe size, ridging and special features such as icebergs. The information

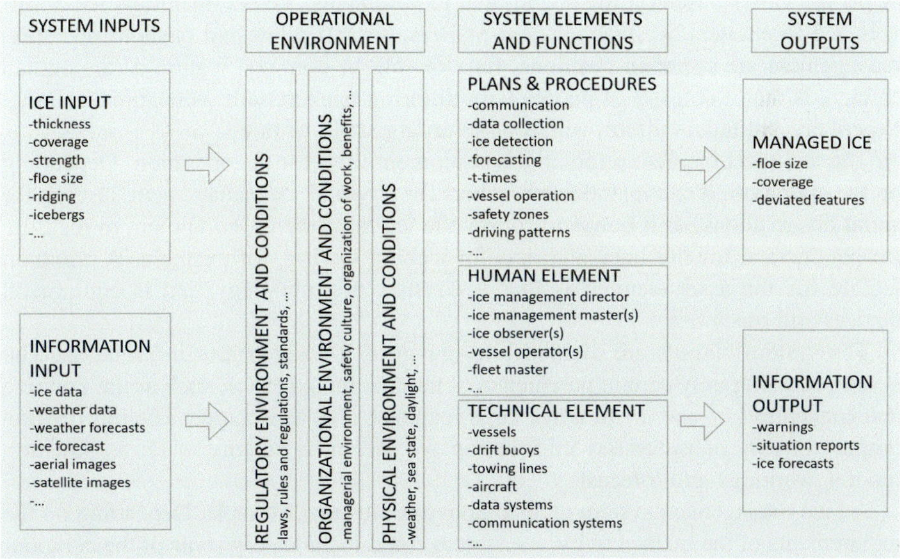

FIGURE 1.10 Ice management principles (Haimelin et al., 2017).

input consists of all information relevant to the operation before processing by the system, such as ice and weather data and forecasts, drift buoy data, aerial and satellite images, for example.

The regulatory environment and conditions component is formed by international and national laws, industry rules, regulations and standards for marine and offshore operations. The type of main activity can also introduce an additional regulatory environment, such as the oil and gas industry standards. For example, the IMO Polar Code, established on 1 January 2017, sets additional training requirements for officers in charge of navigation in polar waters.

The organizational environment and conditions, such as safety culture, company policies, compensations, manning, and organization of work affect the employees' attitudes and the quality of equipment and its maintenance. The safety culture of the organization affects the planning of operations and instructions for procedures.

Ice management operations are conducted in different physical environments and conditions. Depending on the location and the time of the year, the weather and sea conditions can differ greatly. High wind, waves and cold temperature affect equipment, its operability and people. Lack of daylight, fog and rain affect visibility and can affect people's vigilance.

Systematic and safe operations require plans and procedures, covering topics such as communication, data collection, driving patterns, safety zones, responsibilities and command structure. In addition to sea ice, the ice management system processes a large amount of information. A common feature of the plans and procedures element is that it is based on documents and that it relies heavily on information and communications technology (ICT).

The human element refers to all personnel involved in an operation. The number of people varies based on the operational requirements. Vessel operators, ice observers, ice forecasters, ice management masters, aircraft pilots and onshore operation management are common positions, for example. A common feature of the human element is that it consists of people with differing backgrounds, education, training, experience, attitude, culture, values and varying states of mind.

The technical element refers to the equipment used for the operation. Depending on the operation, it can include icebreakers for physical ice management, aircraft for aerial observations, drift buoys to follow the ice movements, equipment to measure ice parameters, towing lines and nets for iceberg towing, for example. A common feature for the asset element is that it consists of technology that is equipment, devices and machines.

The system outputs are divided into managed ice and output information. The system can alter only certain parameters of the incoming sea ice, such as the floe size and concentration, and it can move large features, such as icebergs. The information output consists of processed information for the main activity, such as situation reports, warnings and forecasts.

An ice management system defined above is a typical example. Depending on the requirements of the main activity, the system could include only some of the components or additional ones. For example, when the activity is prepared to disconnect by approaching ice and not to rely on physical ice management, the system could consist only of ice monitoring and detection.

1.4.3 Summary

The volume of Arctic shipping is on the rise, driven by multiple factors. Arctic holds some of the world's largest remaining oil and gas reserves. In recent years, the extraction of these natural resources has resulted in increased destination-Arctic shipping of cargo in and out of the Arctic. This development is likely to continue even though non-fossil fuel developments has been fast during the recent years, but it will take at least 10–15 years before we have a good solution outside fossil fuels for maritime transport.

It is also expected that the in trans-Arctic shipping will increase, especially via the Northern Sea Route (NSR) that offers a significantly shorter distance between Northern Europe and the Far East than the conventional route through the Suez channel. In addition to the expected increase in transport, there is an expected increase in Arctic cruises and tourism. These developments are driven by multiple drivers including climate change, technological development as well as national and international policies.

The ice extent is decreasing both on the Arctic and Antarctic waters. This means that the shipping activities on these waters will increase as there is less ice and it will be easier to navigate in these waters in future. There is a possibility of an ice-free Arctic Ocean for a short period in summer in future earlier than mid-century. This would mean the disappearance of multi-year ice, as no sea ice would survive the summer melt season. It is highly plausible that there will be greater marine access and longer seasons of navigation, except perhaps during winter, but not necessarily less difficult ice conditions for marine operations.

Various ship types have different operation requirements and typically the ship types are divided into three categories: icebreakers, ice-going ships and ice-strengthened ships. Icebreakers have to be able to navigate independently on all the planned ice conditions and they should never get stuck in ice. The hull has to be strong enough for all the planned ice conditions. Ice-going ships can navigate independently in ice on the planned operation area. Ice-strengthened ships are competing with no ice-strengthened ships in open water as most of the time they navigate in open water, therefore, they will have compromised open water and ice-going performance. They can be capable of breaking thin ice but typically they need icebreaker assistance on normal operation conditions.

The principles of icebreaker assistance and ice management are presented, which are two typical important operations in ice. For both of these, the system-level understanding of the ice conditions, ice dynamics, ship performance in ice are important and both of these topics are under active research today.

REFERENCES

AMSA, Arctic Marine Council, 2009. Arctic marine shipping assessment report, April 2009, second printing.

Bambulyak, A., Polach, R. U. F. V. B. U., Ehlers, S., and Sydnes, A., 2014. Challenges with oil spill risk assessment in arctic regions: Shipping along the northern sea route. In *Proceedings of the ASME 2014 33rd International Conference on Ocean, Offshore and Arctic Engineering*, San Francisco, CA, V04BT02A031.

Bergström, M., 2017. *Autonomous in the Arctic – fortune or folly?*, ABB Generations, Marine and Ports, pp. 58–61.

Bergström, M., Leira, B. J., and Kujala, P., 2020. Future scenarios for Arctic shipping. In *International Conference on Offshore Mechanics and Arctic Engineering*. American Society of Mechanical Engineers, V007T07A006.

Critic, C., 2020. New expedition and adventure ships on order [Online]. [15.12.2019].

Crystal Serenity, 2016. www.crystalcruises.com

Eik, K., 2008. Review of experiences within ice and iceberg management. *Journal of Navigation*, 61, 557–572.

Gautier, D. L., Bird, K. J., Charpentier, R. R., Grantz, A., Houseknecht, D. W., Klett, T. R., Moore, T. E., Pitman, J. K., Schenk, C. J., Schuenemeyer, J. H., Sørensen, K., Tennyson, M. E., Valin, Z. C., and Wandrey, C. J., 2009. Assessment of undiscovered oil and gas in the arctic. *Science*, 324, 1175–1179.

Haimelin, R., Goerlandt, F., Kujala, P., and Veitch, B., 2017. Implications of novel risk perspectives for ice management operations. *Cold Regions Science and Technology*, 133, 82–93.

Hamilton, J., Holub, C., Blunt, J., and Mitchell, D., 2011. Ice management for support of arctic floating operations. In *OTC Arctic Technology Conference*, Houston, TX. OTC-22105-MS.

Katona, V., 2017. Low oil prices won't deter russia from arctic drilling [Online]. OilPrice.com. [15.12.2019].

Keinonen, A. J., Browne, R., Lam, E., and Jolles, W., 2006b. Six years in Sakhalin Offshore oil – Management of risk, operations in ice. In *SNAME 7th International Conference and Exhibition on Performance of Ships and Structures in Ice*, Banff, Alberta, Canada.

Keinonen, A., Shirley, M., Liljeström, G., and Pilkington, R., 2006a. Transit and stationary coring operations in the central polar pack. In *SNAME 7th International Conference and Exhibition on Performance of Ships and Structures in Ice*, Banff, Alberta, Canada, July 2006.

Keinonen, A., Wells, H., Dunderdale, P., Pilkington, R., Miller, G., and Brovin, A., 2000. Dynamic positioning operation in ice. In *The Tenth International Offshore and Polar Engineering Conference*, Seattle, USA, May 27–June 2, 2000.

Kubat, I., and Sayed, M., 2014. Literature survey of station keeping and ice management. In *SNAME 11th International Conference and Exhibition on Performance of Ships and Structures in Ice*, Banff, Alberta, Canada. SNAME-ICETECH-2014-148.

Kujala, P., Kulkarni, K., Kondratenko, A., Lu, L., Casper, W., Li, F., and Mashrura, M., 2024. System level simulation of the winter navigation in the Baltic Sea. In *Proceedings of 15th International Marine Design Conference (IMDC-2024)*, Amsterdam, the Netherlands.

Kulkarni, K., Kujala, P., Musharraf, M., and Rainio, I., 2022b. Simulation tool for winter navigation decision support in the Baltic Sea. *Applied Sciences*, 12, 7568.

Kulkarni, K., Li, F., Kondratenko, A. A., and Kujala, P., 2024. A voyage-level ship performance modelling approach for the simulation of the Finnish-Swedish winter navigation system. *Ocean Engineering*, 295, 116997.

Kulkarni, K., Li, F., Liu, C., Musharraf, M., and Kujala, P., 2022a. System-level simulation of maritime traffic in northern Baltic Sea. In *Proceedings of the 2022 Winter Simulation Conference*, Marina Bay Sands, Singapore, December 11–14, 2022.

Matthews, S., Johannessen, O. M., and Michelsen, L.-H. P., 2019. *The geopolitical implications of arctic sea ICE melt*. The Norwegian Climate Foundation.

Melia, N., Haines, K., and Hawkins, E., 2017. Future of the sea: Implications from opening arctic sea routes. *Foresight—Future of the Sea Evidence Review*. UK government's Foresight Future of the Sea project. July 2017, 38 p.

National Snow and Ice Data Center, n.d. Sea Ice Today Analyses and daily images of sea ice conditions.

Topaj, A., and Tarovik, O., 2023. Operational planning and combinatorial optimization in simulation models of marine transportation systems. In *Proceedings of the 25th International Conference on Harbor, Maritime and Multimodal Logistic Modeling & Simulation*. Athens, Greece, September 18–20, 2023.

2 Ice Formation and Properties

2.1 DEVELOPMENT OF SEA ICE

2.1.1 ICE FORMATION PROCESS

Sea ice is a complex material formed when sea water freezes at low temperatures. It consists of solid ice, brine, gas and various solid salts, and is influenced by temperature (Petrich and Eicken, 2016). The formation of sea ice involves four stages (as show in Figure 2.1):

1. As the sea surface temperature drops below the freezing point, condensation nuclei form in the sea water, leading to the gradual formation of small thin ice section.
2. With continued cooling, small thin ice section accumulates and are stacked by winds, currents, waves and tides.
3. Further freezing creates ice crusts or ice cakes, also known as pancake ice, floating on the sea surface.
4. As the temperatures drop further, pancake ice covers the sea surface and thickens, forming grey ice and white ice. Movement of sea water cause continuous fracturing, accumulation and extrusion of ice, along with multiple melting and refreezing of sea ice, resulting in large ice formations.

The formation and development of sea ice are influenced not only by temperature but also by sea water density, salinity, depth, turbulence and condensation nuclei. The density of sea water increases with the decrease of temperature. When surface water cools, its density becomes greater than the water below, causing the cooled water to sink, forming convection. Due to convection, more water needs to be cooled before cooling, thus delaying the time to freeze. During ice formation, brine is trapped in ice crystals and small thin ice sections, while some brine precipitates and sinks, increasing the salinity of lower seawater and intensifying convection.

The water depth and ocean turbulence significantly affect sea ice formation. Fully developed turbulence in shallow waters and the ocean's mixed layer stabilizes the sea water temperature, reducing the cooling effect of surface cooling and delaying ice formation. Especially in autumn and winter, turbulence in the ocean boundary layer increases the thickness of the mixed layer, directly influencing early winter sea ice formation.

The effect of condensation nuclei on sea ice formation is significant. Seawater near the freezing point freezes quickly when it contains many condensation nuclei. In the seas with few or no condensation nuclei, the water remains unfrozen even

DOI: 10.1201/9781003676829-2

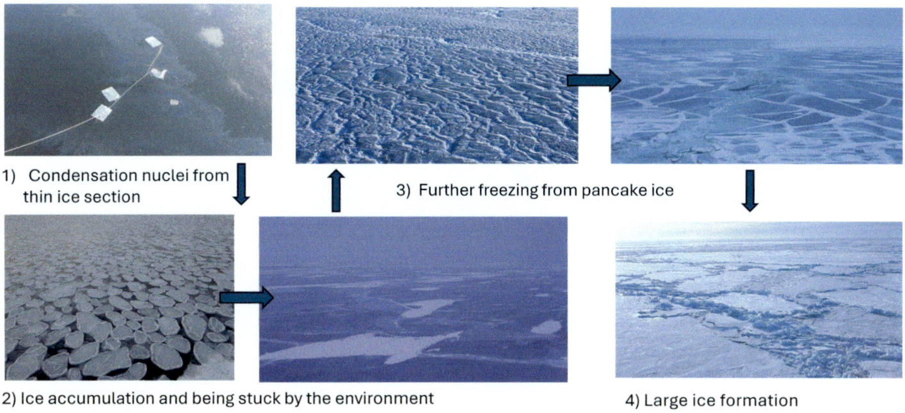

FIGURE 2.1 Ice growth process.

when cooled. However, once ice crystals form, they act as condensation nuclei, causing the sea water freezes quickly.

2.1.2 Ice Thickness Growth

Thickness is the most intuitive characteristic of sea ice. Generally, thicker sea ice tends to be stronger, with higher bearing capacity and ice load on marine structures. As mentioned before, sea ice comes in various forms, but from an engineering perspective, the thickness of level ice is of particular importance. First-year level ice can reach thicknesses of up to 2 metres, while multiyear level ice can be even thicker.

To predict the thickness of sea ice, scientists have devoted considerable effort to simulating the growth of sea ice. Sea ice growth is an extremely complex process full of uncertainties, as it is influenced by freezing time, environment temperature, snow cover, wind speed, thermal radiation and many other factors. Josef Stefan gave already in 1891 (see e.g. Crepeau, 2007) an analytical expression for sea ice thickness based on heat balance, and the following is a simplified form without snow (Yen, 1981):

$$h^2 = \frac{2\lambda_{si}}{\rho_i L}\int_t T_b - T_a dt = \frac{2\lambda_{si}}{\rho_i L}(T_b - T_a)t \tag{2.1}$$

where h is sea ice thickness, λ_{si} is the thermal conductivities of sea ice, ρ_i is the density of sea ice, L is the latent heat coefficient of sea ice, T_b is the temperature at the bottom of sea ice, T_a is the temperature of air, t is freezing time in s.

There is a limit value for first-year sea ice thickness, which can be calculated by variation of Stefan's formula (Thomas, 2017):

$$H_{eq} = -\lambda_{si}\frac{T_a - T_b}{F_w}\left(1 + \frac{\lambda_{si}}{\lambda_s}r_s\right)^{-1} \tag{2.2}$$

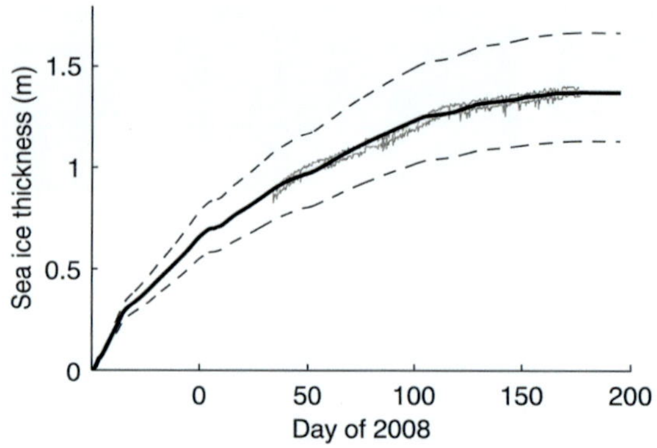

FIGURE 2.2 Sea ice thickness as a function of time-based on a degree-day model (thick line) and estimates of thickness development without snow and with heavy snow load (dashed lines) (Thomas, 2017).

where H_{eq} is the equilibrium thickness of sea ice, λ_s is the thermal conductivities of snow, F_w is the oceanic heat flux from the underlying water into the interface and r_s is the constant of proportionality.

Note that the above formula is based on first-year level ice during the growth season, and may not apply to sea ice under other conditions. Figure 2.2 shows the model's simulation of sea ice growth in the Barrow Sea. As can be seen, the growth rate of sea ice decreases as the ice becomes thicker. Consequently, the differences in thickness between ice floes that begin forming at different times tend to reduce with time.

ISO (19906) (ISO, 2019) recommends a simplified empirical formula for predicting sea ice thickness, taking into account the influence of the air temperature and snow depth:

$$h = -m + \left[(m + h_0)^2 - 405\sum t_a\right]^{1/2} \tag{2.3}$$

where t_a is the mean daily air temperature at the meteorological shelter height, m is an empirical coefficient that depends on snow depth, h_0 is ice thickness at the start of the period.

2.1.3 Crystalline Structure of Sea Ice

The growth process of sea ice determines its crystalline structure. At the initial stage of growth when nuclei forms, the *c*-axes of the ice crystals orient randomly in space, resulting in granular layout of ice crystals. Once a continuous ice layer forms at the surface, ice crystals with different orientations at the ice-water interface compete for further growth. Since an ice crystal grows faster in the directions perpendicular to its *c*-axis rather than along it, crystals with *c*-axes oriented horizontally dominates the downward growth over those with *c*-axes oriented vertically, as illustrated in Figure 2.3. As a result, *c*-axes of ice crystals become aligned and orient randomly in the horizontally plane with further ice growth, resulting in columnar layout of ice crystals.

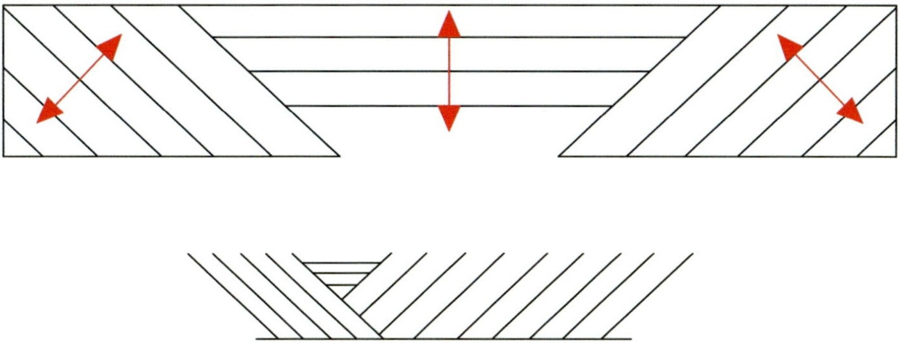

FIGURE 2.3 Inclined axis crystals grow rapidly, causing vertical growth to gradually disappear.

(Reproduced from Weeks and Wettlaufer, 1996.)

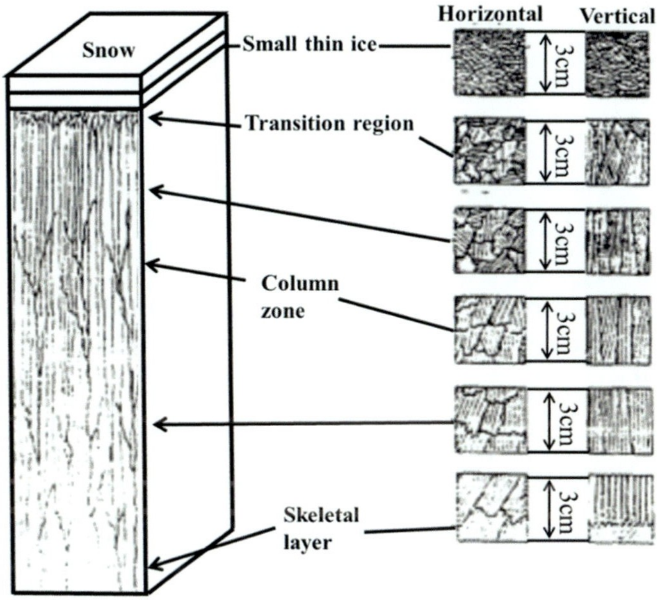

FIGURE 2.4 The structures of first-year ice (Assur, 1960).

The grain structure of sea ice can be observed by optical techniques. This can be done by milling ice into thin sections and putting it in between two polarizers. The crystals will exhibit different colours depending on its orientation of *c*-axes. We can then characterize the type of ice according to the alignment of *c*-axes and grain size. Figure 2.4 presents an example of the observed grain structure of first-year ice.

First-year ice is typically composed of a thin layer of granular ice crystals at the top, followed by a transition zone where crystals are getting aligned, and then a long columnar section further downwards. At the bottom of the ice structure, ice crystals may get more aligned towards one specific direction in the horizontal plane as the result of ocean current. The grain size is smaller in the granular zone and gets larger in the columnar zone downwards. During ice growth, brine and air are trapped in between the crystals, forming brine pocket and air pockets. The brine drains downwards as a result of gravity, forming vertically expanded brine channels. These factors make sea ice porous. The apparent difference of ice structure in vertical and horizontal direction results in transverse isotropic mechanical behaviour of first-year ice.

The structure of old ice (second-year and multiyear ice) differs from that of first-year sea ice, due to the repetitive melting and freezing process as well as the mechanical deformation such as ridging. The vertical section is a mixture of granular and columnar ice crystals with a large variation of grain size; brine and air pockets are rarely seen. As a result, the mechanical property of multiyear ice is often isotropic.

The characterization of sea ice is carried out according to its crystal structure, including the grain size and shape as well as the orientation and alignment of *c*-axes. The component of sea ice can be classified into primary ice and secondary ice. Primary ice is formed as the thin top layer of sea ice, which is granular. It is divided into *P1*, *P2*, *P3* and *P4* ice according to grain size and shape. *P1* has the large grain size and others decrease in turn. Among them, *P3* is the most common in sea ice, with small to medium size grain and random *c*-axes orientation. Secondary ice is formed after primary ice, which is columnar. It is divided into *S1*, *S2*, *S3*, *S4* and *S5* according to grain size and the orientation of *c*-axis. Among them, *S2* ice is the most common type in sea ice. It is featured with *c*-axes randomly oriented in the horizontal plane, and is the main component of first-year sea ice. *S3* type of ice forms at the bottom of thick ice sheets and covers, with *c*-axes aligned to a specific direction (Assur, 1960).

2.2 DEFORMATION AND FORMS OF SEA ICE

Sea ice exists in nature by various forms and may experience significant deformation.

Landfast ice. Sea ice that remains fast along the coast, over shoals or between grounded icebergs (also called *fast ice*, as shown in Figure 2.5). In most areas, landfast ice forms in the autumn, persists throughout the winter and melts in the spring and summer. In some places, it can remain throughout the summer, becoming multiyear ice.

Drift ice/pack ice. Term used in a wide sense to include any sea ice other than fast ice, as shown in Figure 2.6. When the ice concentration is higher than 7/10, pack ice can be used instead.

Pancake ice. Pieces of new ice, usually approximately circular, about 30 cm to 3 m across and with raised rims due to the pieces striking against each other,

FIGURE 2.5 Landfast ice.

FIGURE 2.6 Drift ice.

as shown in Figure 2.7. It primarily consists of circular ice with a diameter of 30 cm to 300 cm and a thickness of 10 cm. These ice floes have raised edges from collisions and can freeze into grease ice or pancake ice. They can also break into ice floes or frazil ice in large wind waves or swell. Sometimes, they form at the interface of water bodies with different physical characteristics at certain depths and then float to the surface, quickly covering large sea areas.

Ice floe. Any relatively flat piece of ice 20 m or more across, as shown in Figure 2.8. Depending on its size, WMO (2014) defines different terms, listed below:

- *Floe small*: With horizontal scale of 20 m~100 m;
- *Floe medium*: With horizontal scale of 100 m~500 m;

FIGURE 2.7 Pancake ice.

FIGURE 2.8 Ice floe.

- *Floe big*: With horizontal scale of 500 m~2000 m;
- *Floe vast*: With horizontal scale of 2 km~10 km;
- *Floe giant*: With horizontal scale more than 10 km.

Level ice. Sea ice that is unaffected by deformation, as shown in Figure 2.9 (a substitute term is *undeformed ice*).

Deformed ice. A general term for ice that has been squeezed together and in places forced upward and downward, as shown in Figure 2.10 (a substitute term is *pressure ice*), including rafted ice, ice ridge, hummock, hummocked ice, etc.

FIGURE 2.9 Level ice.

FIGURE 2.10 Deformed ice field.

Rafted ice. A form of pressure ice in which one floe overrides another, as shown in Figure 2.11. Note that sometimes even three or four layers of ice overlap each other, but the angle of the overlapping surface is not large, and the ice is still relatively flat.

Ice ridge. A ridge or wall of broken ice forced up by pressure. The ice ridge can be divided into three parts:

- Ridge sail: Above the consolidated layer as shown in the Figure 2.12.
- Consolidation layer: Above the waterline and below the ridge sail, the consolidation layer is two to three times thicker than the surrounding layer of ice.
- Ridge keel: The parts below the consolidation layer, it consisted of broken ice.

FIGURE 2.11 Rafted ice.

FIGURE 2.12 Ice ridge.

Brash ice. Accumulations of ice made up of fragments no more than 2 m across, as shown in Figure 2.13 (the wreckage of other forms of ice).

Iceberg. A large piece of ice of different shapes that has escaped from a glacier or ice shelf and is floating on the sea surface or stranded on land, rising more than 5 m above sea level, as shown in Figure 2.14. Depending on its size, iceberg can be categorized into glacial berg, bergy bit, growler, etc.

Sea ice concentration refers to the coverage of ice groups on the sea surface, which is the ratio of the covered area of sea ice to the total area of the sea, indicating the coverage degree of floating ice in the sea area, usually using the ten-point method, as shown in Table 2.1.

FIGURE 2.13 Brash ice channel.

FIGURE 2.14 Iceberg.

TABLE 2.1
Ice Concentration Definition (WMO, 2014)

Ice Concentration	Definition
0/10	Ice Free
<1/10	Open Water
1/10~3/10	Very Open Pack Ice
4/10~6/10	Open Pack Ice
7/10~8/10	Close Pack Ice
9/10	Very Close Pack Ice
10/10	Consolidated pack ice compact

2.3 ICE PHYSICAL PROPERTIES

The complex growth environment and historical development of sea ice significantly influence its diverse physical properties. Among these, the most prominent physical characteristics are temperature, density, salinity and porosity.

2.3.1 Temperature

Temperature is a fundamental property of sea ice that significantly influences its grain structure, the phase state of impurities, and consequently, other physical and mechanical properties of the ice. When measuring these physical properties, ice core samples are collected from the site, and the temperature of the ice is determined by drilling a hole and inserting a thermistor into the core. An example of a typical sea ice temperature profile is shown in Figure 2.15.

It is evident that the temperature distribution within sea ice is not uniform, particularly in polar regions. For first-year sea ice during the growth season, the temperature distribution can generally be assumed to be linear. However, in the case of multiyear sea ice, which has undergone melting season, the temperature profile becomes more complex and variable, and thus is not discussed here.

The temperature at the lower surface of sea ice can be considered equivalent to the freezing point of seawater, which itself varies depending on the salinity of the seawater. In contrast, the temperature at the upper surface of sea ice is influenced by several factors, including air temperature, snow cover, wind speed, and other environmental conditions, and it satisfies energy balance. Cox and Weeks (1988) formulated an expression for the energy balance at the upper surface of sea ice. Readers interested in this expression can refer to the original source.

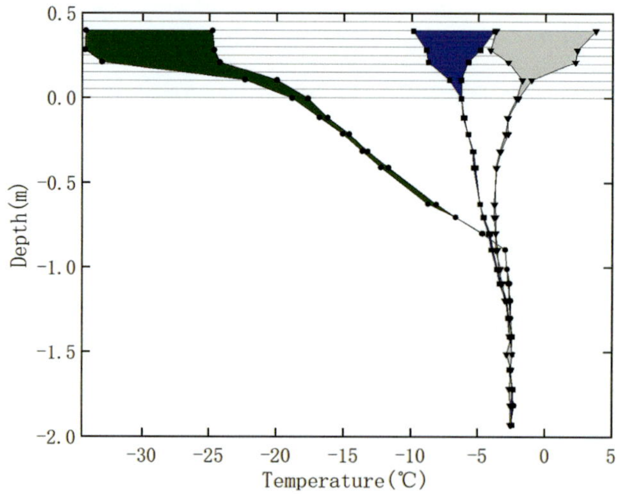

FIGURE 2.15 Temperature profiles of sea ice.

(re-produced from Thomas, 2017). Positive depths are snow and air, while negative depths are sea ice and ocean. Profiles show the temperature range encountered over a 24-hour period.

2.3.2 Density

Determining the density of sea ice is challenging due to the presence of impurities such as solid salts and trapped gases. Currently, four primary techniques are employed for density measurements: mass/volume, displacement (submersion), specific gravity, and freeboard and ice thickness methods. These measurements are typically conducted in the field, in order to ensure the in-situ density of the sea ice.

Timco and Frederking (1996) summarized field data from 21 investigations, showing that the density of first-year sea ice ranges from 0.75 Mg m^{-3} to 0.96 Mg m^{-3}, while the density of multiyear ice varies from 0.72 Mg m^{-3} to 0.91 Mg m^{-3}. The density of sea ice generally decreases with age due to the gradual drainage of brine. Additionally, the density of ice above the waterline is lower than below the waterline, as brine drains from the above-water portion of the sea ice, which is then replaced by air.

A theoretical method was proposed by Cox and Weeks (1983) to calculate the density of sea ice based on the salinity and temperature of sea ice:

$$\rho = (1 - v_a)\frac{\rho_i F_1(T)}{F_1(T) - \rho_i S_i F_2(T)} \tag{2.4}$$

where ρ is the density of sea ice, v_a is the porosity of gas, ρ_i is the density of pure ice, S_i is the salinity of sea ice, $F_1(T)$ and $F_2(T)$ are functions of temperature.

Anderson (1960) provides curves illustrating the relationship between density, temperature and salinity of gas-free ice, as shown in Figure 2.16. These curves clearly demonstrate that sea ice density generally increases with rising salinity. Additionally, density is highly sensitive to temperature near the freezing point

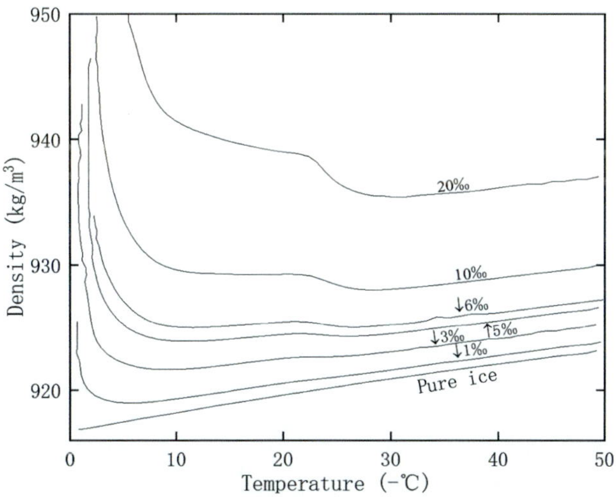

FIGURE 2.16 Density of gas-free sea ice as a function of temperature and salinity.

(Re-produced from Anderson, 1960.)

but is less affected at temperatures further from this point. Notably, the curves exhibit a decline around −23°C, which occurs due to the phase transition of $NaCl{\cdot}2H_2O$ salts from liquid to solid.

2.3.3 Salinity

The grains of sea ice are inherently pure and free of salt, but during the growth of sea ice, some seawater becomes trapped between the grain layers, resulting in the presence of salinity within the ice. Generally, sea ice that forms rapidly tends to trap more brine and thus exhibits higher salinity, whereas slow-growing sea ice contains fewer impurities. The salinity of sea ice is determined by melting a sample of the ice and measuring the salinity of the saltwater. The critical aspect of this measurement process is preserving the sample to prevent brine from draining out and to ensure that the salinity remains consistent. Salinity is typically expressed in parts per thousand (ppt, or procentual).

As sea ice continues to grow, brine is expelled from the ice through two primary mechanisms: brine expulsion and gravity drainage. Brine expulsion occurs when the temperature of sea ice drops, causing the water within the brine to freeze and thereby concentrating the remaining brine. Since the volume of ice is greater than that of an equivalent mass of water, the excess brine is forced out of the ice. Gravity drainage, on the other hand, occurs because the brine in the colder upper layers is denser than the brine in the warmer lower layers. This density difference drives convective overturning, resulting in the drainage of brine from the bottom of the sea ice.

To estimate the average salinity (S_i) of sea ice during the growing season, Ryvlin (1974) proposed a formula that incorporates several influencing factors, including seawater salinity (S_w), ice thickness (h_i), rate of ice freezing (a) and final ice/seawater salinity ratio (b):

$$S_i = S_w \cdot \left(\frac{1-b}{e^{10 \cdot a \cdot \sqrt{h_i}}} + b \right) \tag{2.5}$$

Cox and Weeks (1974) collected extensive data on the average salinity of first-year sea ice in both the Arctic and Antarctic regions and developed a fitted formula, which is also recommended by ISO 19906:

$$\begin{cases} S_i = 13.4 - 17.4h_i, h_i \leq 0.34m \\ S_i = 8.0 - 1.62h_i, h_i > 0.34m \end{cases} \tag{2.6}$$

The equations mentioned are plotted in Figure 2.17. The findings indicate that the salinity of first-year sea ice in polar regions is strongly dependent on ice thickness. Notably, at an ice thickness of approximately 0.34 meters, the slope of the curve decreases significantly, which is believed to result from a shift in the brine drainage mechanism from brine expulsion to gravity drainage.

It is important to note that these equations do not apply to the warming season or multiyear ice, as the salinity of sea ice after melting no longer maintains a positive correlation with the growth period. In the case of multiyear ice, most of the salt has drained out, resulting in a typical salinity range of 0.5–4.0 ppt.

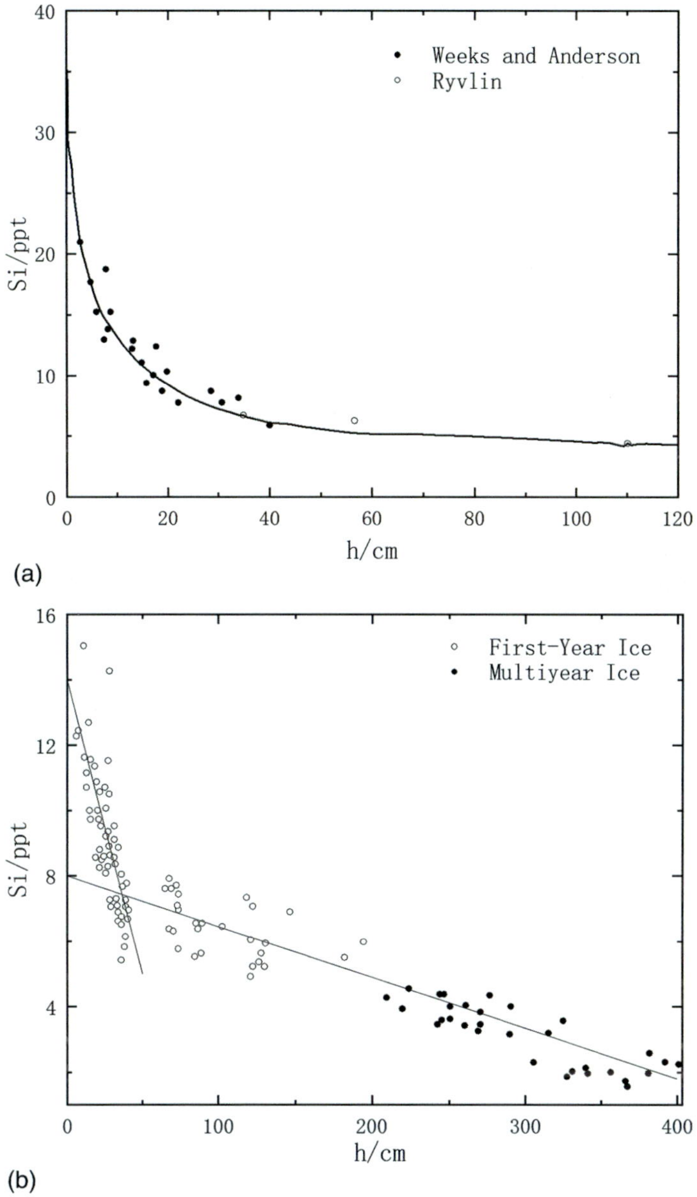

FIGURE 2.17 Estimating formula of average salinity of sea ice during the growth season. **(Re-produced from Ryvlin, 1974; Cox and Weeks, 1974.)**

As mentioned before, the salinity of sea ice is not uniformly distributed due to its growth processes and brine drainage mechanisms. Malmgren (1927) investigated the evolution of salinity profiles in Arctic first-year ice throughout the winter and into the summer season, as illustrated in Figure 2.18. During the growth season, the salinity profiles of first-year sea ice typically exhibit a C-shaped pattern. However, for melted sea ice (from June to August), the salinity profiles become more variable.

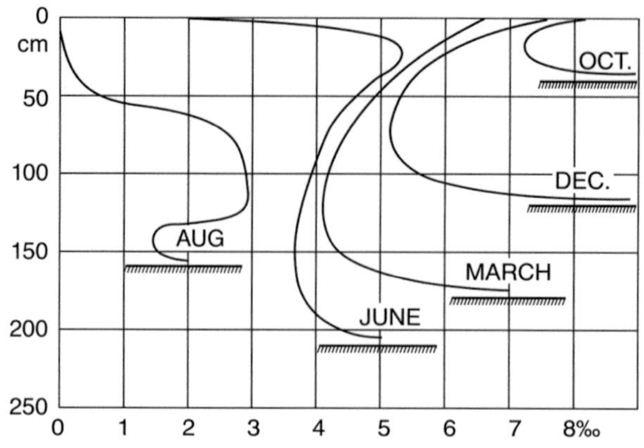

FIGURE 2.18 Evolution of first-year sea ice salinity profiles (Malmgren, 1927).

Cox and Weeks (1988) developed a method to estimate salinity profiles for first-year sea ice during the growth season, and the results are in good agreement with the field data. Detailed introduction is not given here, interested readers can refer to the original literature.

2.3.4 Porosity

In engineering, porosity is commonly more used to estimate the properties of sea ice than salinity. Porosity refers to the ratio of the void volume to the total volume of sea ice, where the voids may contain brine or gas (referred to as brine porosity and air porosity, respectively). The unit of porosity is expressed in parts per thousand (ppt), similar to salinity.

Cox and Weeks (1983) gave the theoretical formula of brine volume:

$$v_b = \frac{\rho S_\mathrm{i}}{\rho_b S_b (1 + k)} \tag{2.7}$$

where v_b is the brine volume, ρ_b is the brine density, S_b is the salinity of brine, k is the coefficient related to the solid salt content determining by Assur's (1960) phase equilibrium table.

ISO 19906 recommends a better way to estimate brine volume, which can be simply calculated based on sea ice salinity and temperature:

$$v_b = S_\mathrm{i}\left(0.532 + \frac{49.185}{|T_i|}\right), \; -0.5^\circ\mathrm{C} \geq T_i \geq -22.9^\circ\mathrm{C} \tag{2.8}$$

Cox and Weeks (1988) also simulated brine volume profile, as shown in Figure 2.19, and were surprised to find that it differed from the sea ice salinity profile. They discovered that the brine volume profile was solely a function of ice thickness and was independent of growth data.

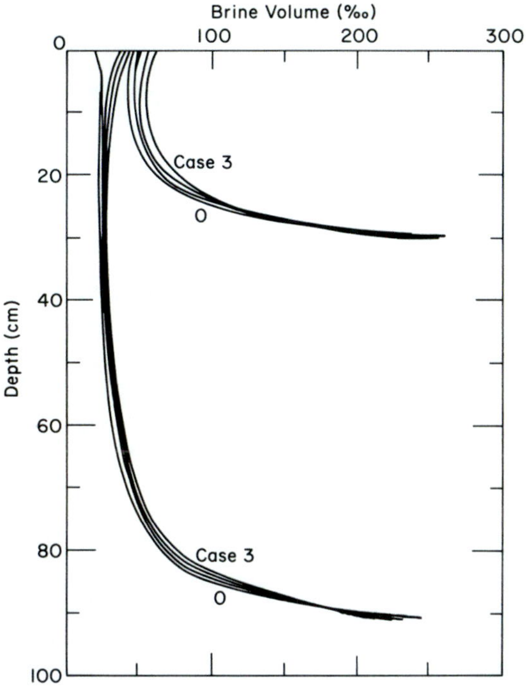

FIGURE 2.19 Brine volume profiles for different thickness and temperature (Cox and Weeks, 1988).

In the previous section, we discussed how brine gradually drains from sea ice as it grows. As this drainage occurs, air replaces the original brine volume, which is the primary cause of air porosity. Consequently, it is reasonable to disregard air porosity in new or young ice. However, in older ice that has lost a significant amount of brine, the impact of air porosity becomes substantial and cannot be ignored.

The air volume within sea ice is typically measured by melting the ice and capturing the volume of gas released. Cox and Weeks (1983) provided a theoretical formula for calculating air volume based on the density of the ice:

$$v_a = 1 - \rho \cdot \left(\frac{1}{\rho_i} - \frac{F_2(T)}{F_1(T)} \cdot S_i \right) \tag{2.9}$$

As previously mentioned, air porosity is closely related to brine porosity. Tarovik (2022) identified the following relationship between the total porosity (v_t) and brine volume based on field data:

$$v_t = v_b + v_a = 0.871 v_b + 29 \tag{2.10}$$

2.4 ICE MECHANICAL PROPERTIES

The grain structure and the physical properties of sea ice together determine its mechanical properties, including at least tensile strength, shear strength, flexural strength, uniaxial and multiaxial compressive strength, strain modulus, Poisson's ratio and fracture toughness.

2.4.1 Ice as an Engineering Material

Sea ice is a very complex material formed under natural conditions, with varying structure, impurities and temperature across its thickness. As a result, sea ice normally behaves as a transverse isotropic material, with properties remaining the same in the horizontal plane but varying across the thickness. The variation is then the joint effect of temperature, grain structure and porosity, which originates from the ice formation process involving sea water salinity, sea states and temperature variations.

In the engineering world, ice often exists at air temperature close to or even higher than its melting point. Ice is hence a warm material compared to other engineering materials such as steel. As a result, ice exhibits viscosity behaviour, and the deformation process is sensitive to the strain rate. Besides, the mechanical properties of ice are rather sensitive to temperature, especially when it is near the melting point.

Under low strain rate (i.e., slow loading), ice exhibits ductile behaviour with creep as the dominating failure mode. Under high strain rate (i.e., fast loading), ice behave brittlely and the failure is dominated by fracture. This is schematically shown in Figure 2.20. Typically, the interaction between ships and ice is relatively fast and the strain rate is above $10^{-3}\ s^{-1}$. Therefore, ice is often regarded as a brittle material when it comes to ice-going ships. However, as introduced later in this section, ductile behaviour can take place in the ice field when the confinement is high at complex stress state.

Another typical feature of ice mechanical properties is the high randomness. There can be several times difference between the maxima and the minima, even if the mechanical tests are carried out at the same location. This is a common feature among brittle materials as the failure is dominated by fracture, which is eventually determined by the initial flaw in the material. The random distribution of brine pockets and air pockets in ice acts as stress concentrators when ice is loaded. Cracks are then initiated and propagates when the stress intensity reaches the critical level.

2.4.2 Tensile and Flexural Loading

Tensile failure is a dominating failure mode when ice interacts with ships and offshore structures. However, ice in the engineering world is rarely loaded under pure tension. For example, when ships break ice, tensile failure mostly occurs in the form of flexural failure with ice bent downwards, where the upper layer of ice is under tensile state. In field tests, direct measurement of tensile stress is much less frequently carried out compared to indirect measurement through bending, due to the difficulty in test set-ups and sample machining. It is, therefore, common to use flexural strength obtained from bending test to approximate the tensile strength of ice.

Bending tests are usually carried out by cantilever beam tests and three-point bending tests. Cantilever beam tests are carried out in situ without taking the ice sample out of the water. The advantage is that this mimics the most real condition as when ships break ice. Nonetheless, cantilever beam tests typically suffer from stress concentration issues at the root of the beam, resulting in underestimation of ice flexural strength. In addition, bending tests assume homogeneous property across the thickness, which apparently do not hold for sheet ice. The flexural strength measured by cantilever beam tests is then an engineering index of the overall ice-bearing capacity, rather than a material property. Cantilever beam tests also suffer from practical difficulties when ice gets thick, as it requires rather long beams to be prepared and large force to break it. These drawbacks can be overcome by three-point bending test, which is the most common way to carry out ice flexural strength test in polar ship engineering. With three-point bending tests, ice samples are taken out of the water and machined into smaller size before being loaded. It is then possible to load different layers of ice through thickness separately. The drawback lies in the ex-situ test which may change the temperature and brine volume compared to its original state.

Other methods to determine the tensile or flexural strength of sea ice also exist. Karulina et al. (2019) proposed to use Y-shaped cantilever beams to overcome the stress concentration issue in conventional cantilever beam tests, which is shown to be effective. Makarov et al. (2023) conducted Brazilian tests to measure tensile strength of sea ice and concluded that this is a much easier, yet effective, method compared to direct tensile strength measurement. In-situ four-point bending tests have been carried out as the alternative to ex-situ three-point bending, e.g., by Kujala et al. (1990), Taylor et al. (2021).

The flexural strength of first-year sea ice is typically in the range of 0.1–1 MPa. For multiyear ice the flexural strength can be even higher. Flexural strength of sea ice depends on multiple factors, including at least temperature, porosity and grain size. Flexural strength of sea ice increase with lower temperature, lower porosity and smaller grain size, as exemplified in Figure 2.21. The strain rate has little influence on the flexural strength. ISO 19906 recommended the following equation to calculate the flexural strength of sea ice:

$$\sigma_f = 1.76 \exp\left(-5.88\left(\frac{v_b}{1000}\right)^{0.5}\right) \tag{2.11}$$

where σ_f is the flexural strength and v_b the brine volume. The effect of temperature is implicitly embedded into the brine volume, as higher temperature increases brine volume.

2.4.3 Uniaxial Compression

When ships break ice, compressive failure always persists. The strength of ice under compression determines the pressure between ships and ice. Although ice in the real engineering world is typically under complex stress state, uniaxial compressive strength can serve as a reference index on the bearing capacity of ice under compression.

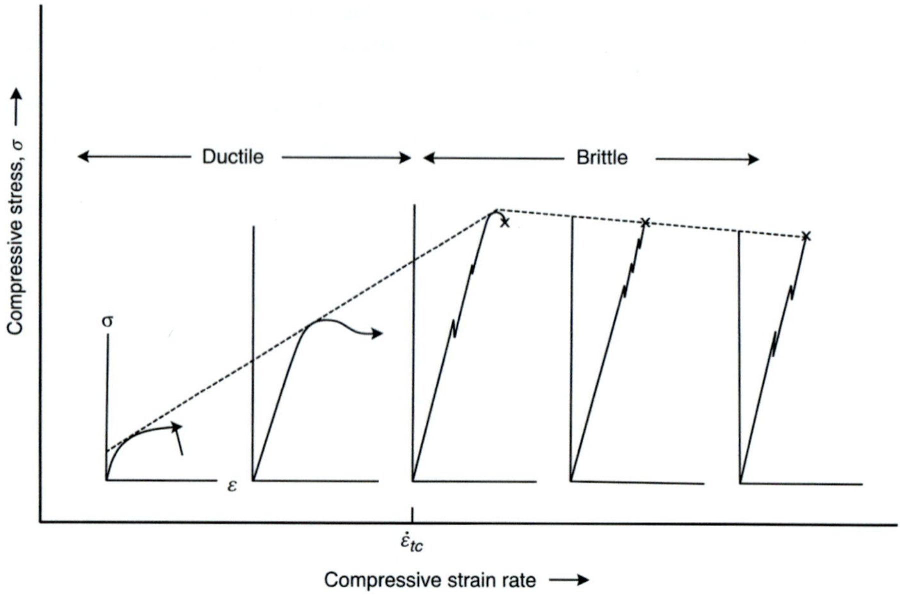

FIGURE 2.20 Ductile-Brittle transition of ice (Schulson and Duval, 2009).

Unlike tensile strength which is difficult to measure directly, it is rather easy to conduct uniaxial compressive tests. This is because ice samples for compressive tests are much smaller than those for tensile tests, and compressing the samples is much easier than pulling them.

Unlike tensile loading, the behaviour of ice under compression depends significantly on the strain rate. As mentioned earlier, ice exhibits ductile behaviour at small strain rate, where the compressive strength increases as a function of strain rate. At high strain rate, ice behaves brittlely and the compressive strength decreases with strain rate further increasing. The compressive strength of first-year ice in the brittle regime is typically in the range of 0.5 MPa–5 MPa, while for multiyear ice, the compressive strength can be even higher (up to 15 MPa). It can be seen that the compressive strength of ice is several times higher than the tensile/flexural strength. This is a common feature of brittle material due to the mechanism of fracture.

In addition to strain rate, other factors including temperature, porosity and grain size also affect the compressive strength of sea ice. Similarly to flexural strength, compressive strength of sea ice increase with lower temperature, lower porosity and smaller grain size, as exemplified in Figure 2.22. It should be noted that the compressive strength differs when ice is loaded horizontally or vertically. Sea ice is usually stronger in vertical direction as the result of its columnar grain structure. For comparison, in bending tests, ice is always loaded in the way that the horizontal plane is under tension, therefore, the flexural strength results represent horizontal loading condition.

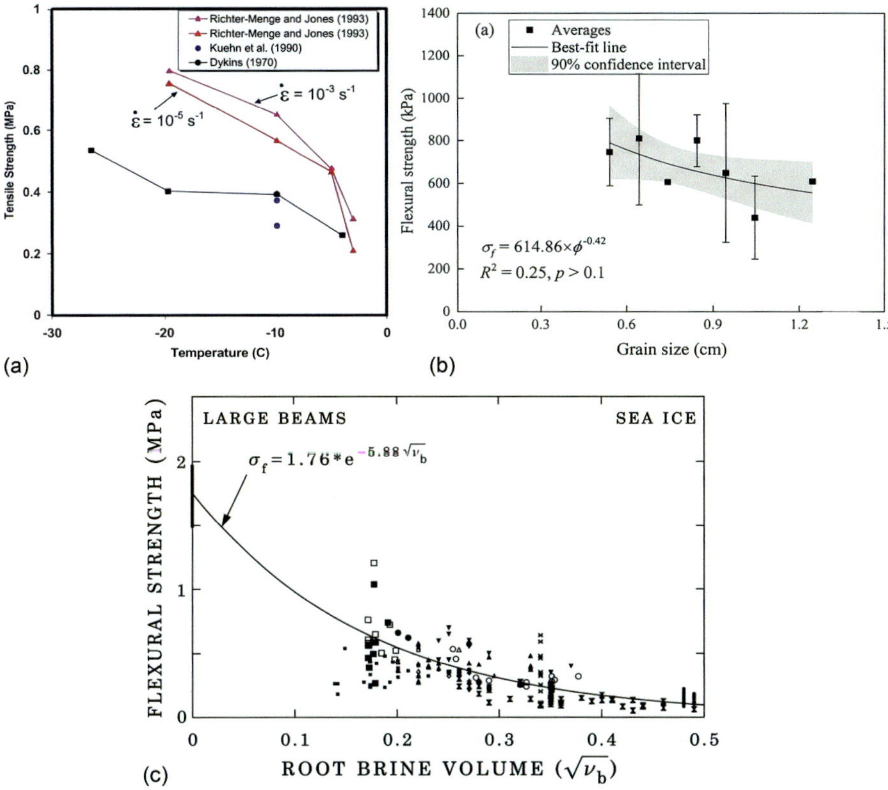

FIGURE 2.21 Flexural strength versus temperature, grain size, and brine volume (Schulson and Duval, 2009; Timco and Weeks, 2010; Wang et al., 2023).

ISO 19906 recommended the following equation to calculate the uniaxial compressive strength of sea ice when it is horizontally loaded:

$$\sigma_c = 37\dot{\varepsilon}^{0.22}\left[1 - \sqrt{\frac{v_T}{270}}\right] \tag{2.12}$$

where $\dot{\varepsilon}$ is the strain rate, v_T is the total volume fraction of brine and air. When vertically loaded, the equation becomes:

$$\sigma_c = 160\dot{\varepsilon}^{0.22}\left[1 - \sqrt{\frac{v_T}{200}}\right] \tag{2.13}$$

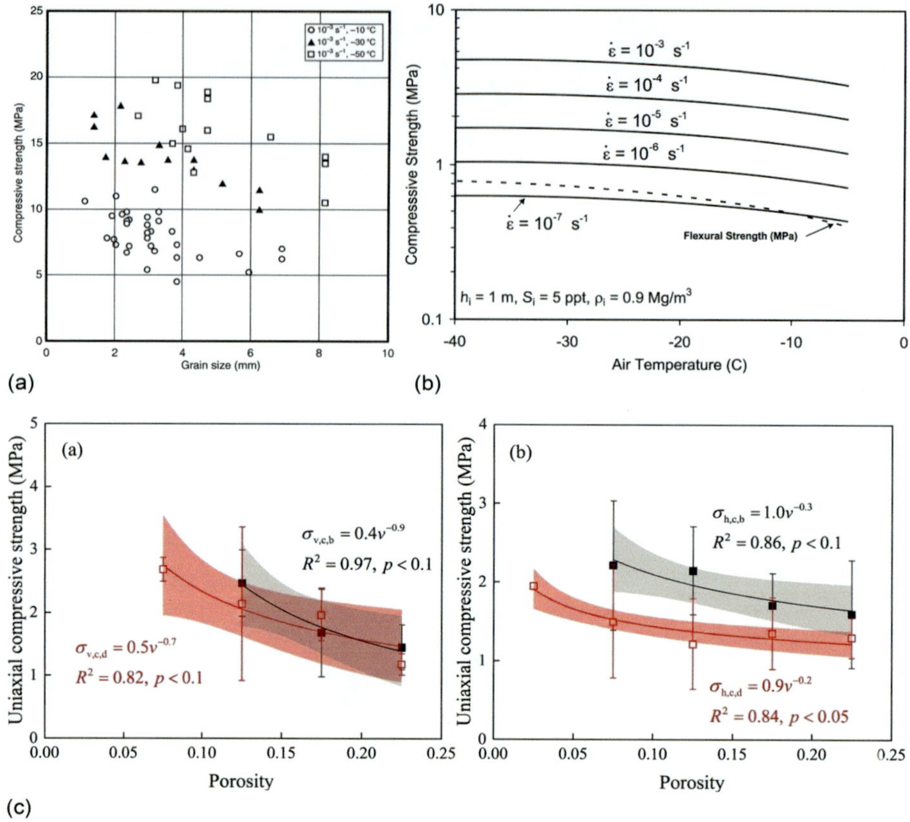

FIGURE 2.22 Compressive strength versus temperature, grain size, and brine volume (Schulson and Duval, 2009; Timco and Weeks, 2010; Wang et al. 2023).

2.4.4 Elastic Modulus and Poisson's Ratio

The ratio of stress and strain during the elastic deformation process is called elastic modulus or Young's modulus. The easiest way to obtain the elastic modulus is to analyse the force-displacement curve of the mechanical tests, such as cantilever beam or three-point bending tests. This approach, however, may underestimate the true elastic modulus due to the viscous behaviour (i.e., the delayed elasticity) of ice. It is therefore more appropriate to refer to the measured elasticity as the *effective elastic modulus*, which is again an engineering index of how ice deforms while loaded by ships. The true elastic modulus can be measured by measuring the propagation of elastic waves, which is more complex. The elastic modulus of sea ice typically locates in the range of 2–10 GPa, and largely depend on the brine volume. ISO 19906 recommended the following equation to calculate the elastic modulus.

$$E = E_0\left(1 - \sqrt{\frac{\nu_T}{1000}}\right)^4 \tag{2.14}$$

The Poisson's ratio defines the ratio between the lateral strain and the longitudinal strain along the loading direction in uniaxial tests. Similar to the elastic modulus, the viscosity affects the measurement so from mechanical tests, the measured quantify is the effective Poisson's ratio. Measurement of Poisson's ratio is much more difficult and hence less frequent compared to other mechanical properties. Therefore, there is still much to discover for Poisson's ratio. ISO 19906 recommended the following equation to calculate the Poisson's ratio.

$$\nu = 0.333 + 0.06105e^{\frac{T_i}{5.48}} \tag{2.15}$$

where T_i denotes ice temperature.

2.4.5 Fracture Toughness

Fracture toughness reflects the capability of a material to resist the propagation of existing cracks. It is an important parameter for brittle materials, as the failure of such materials is dominated by fracture. Nonetheless, the concept of fracture mechanics has not been widely used in the practical design of ice-going ships. Despite that, fracture mechanics does play an important role in the failure of ice, e.g., the splitting of ice floes, and is often used in research related to ships breaking ice.

The mode-I fracture toughness, denoted by K_{IC}, is related to the critical energy release rate, denoted by G_{IC}, through the equation

$$G_{Ic} = \frac{K_{Ic}^2}{E(1-\nu^2)} \tag{2.16}$$

The fracture toughness of sea ice can be measured through standard tests such as four-point beam splitting with a precut notch. Existing measurements shows that the fracture toughness of first-year sea ice is in the order of 10^2 kPa × $m^{0.5}$. The corresponding critical energy release rate is typically in the order of 10^0 J/m^2. There have not been measurements of the fracture toughness of old ice, according to Schulson and Duval (2009).

2.4.6 Complex Stress State and Failure Envelope

In the real engineering world, ice is rarely loaded uniaxially. Instead, the deformation of ice is confined and hence ice is in complex stress states. To understand ice behaviour in such states, biaxial and triaxial compressive tests need to be conducted. On-site measurements of ice compressive stress under multiaxial loading are much fewer compared to uniaxial loading, because of the required procedures involved in such tests. By conducting multiaxial tests, we can get the failure envelope of ice, which denotes the strength of ice under complex stress states. Iliescu and Schulson (2004) compiled some experimental data of sea ice and freshwater as the failure envelope, which is shown in Figure 2.23. As per this figure, when there is no confinement (σ_{22} being zero), ice is under uniaxial loading and the compressive strength is about 5 MPa;

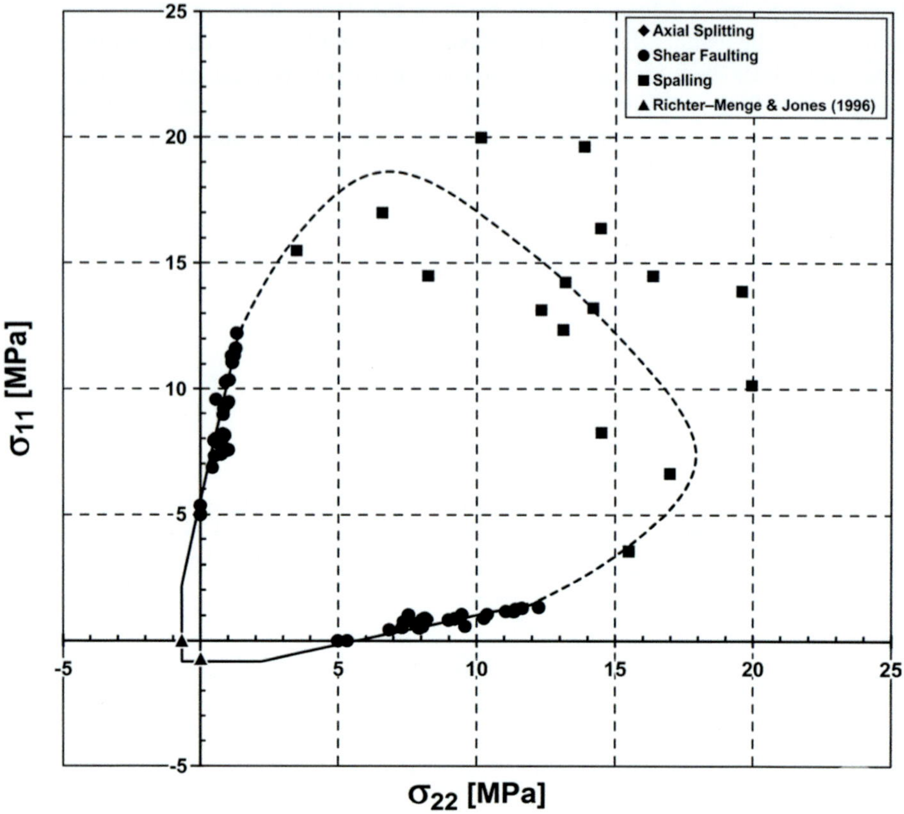

FIGURE 2.23 Failure envelope of first-year sea ice (Schulson et al., 2006).

when the confinement stress gets to 5 MPa, the compressive strength reaches 15–20 MPa, which is over three times of the uniaxial compressive strength.

According to Schulson and Duval (2009), under tri-axial loading with high confinement, the failure of ice may transit from brittle regime to ductile regime, and ice can exhibit plastic behaviour before being damaged. The mechanism of ice under tri-axial loading is very complicated and not fully understood to date. We will not go deeper into this topic. Readers who are interested are referred to Schulson and Duval (2009).

2.5 SUMMARY

This chapter introduces the ice formation process, different forms of ice, as well as the grain structure, physical properties and mechanical behaviour of sea ice. The formation process determines the grain structure and the impurities, which together affects the physical property and mechanical behaviour of sea ice. Sea ice is among the most complicated materials in the engineering world, with still many unknowns to be discovered. Ice engineers should understand the general mechanical behaviour of sea ice and can tell how the properties change at various conditions.

REFERENCES

Anderson, D. L., 1960. *The physical constants of sea ice*. Arctic Institute of North America.

Assur, A., 1960. *Arctic sea ice: Composition of sea ice and its tensile strength*. Washington, DC: National Academy of Sciences-National Research Council Publication.

Cox, G. F., and Weeks, W. F., 1974. Salinity variations in sea ice. *Journal of Glaciology*, 13(67), 109–120.

Cox, G. F., and Weeks, W. F., 1983. Equations for determining the gas and brine volumes in sea-ice samples. *Journal of Glaciology*, 29(102), 306–316.

Cox, G. F. N., and Weeks, W. F., 1988. Numerical simulations of the profile properties of undeformed first-year sea ice during the growth season. *Journal of Geophysical Research: Oceans*, 93(C10), 12449–12460.

Crepeau, J., 2007. Josef Stefan: His life and legacy in the thermal sciences. *Experimental Thermal and Fluid Science* 31 (7): 795–803.

Iliescu, D., and Schulson, E. M., 2004. The brittle compressive failure of fresh-water columnar ice loaded biaxially. *Acta Materialia*, 52(20), 5723–5735.

ISO, 2019. Petroleum and natural gas industries — Arctic offshore structures. ISO 19906. https://www.iso.org/standard/65477.html [18.03.2026]

Karulina, E. B., Marchenko, A. V., Sakharov, A. N., Karulina, M. M., Chistyakov, P. V., and Onishchenko, D. A., 2019. Features of determining the ice flexural strength and the elastic modulus based on floating cantilever beam tests. In *Proceedings of the 25th International Conference on Port and Ocean Engineering under Arctic Conditions*, Vol. 592, pp. 9–13.

Kujala, P., Riska, K., Varsta, P., Koskivaara, R., and Nyman, T., 1990. Results from in situ four point bending tests with Baltic Sea ice. In *Proceedings of the IAHR Symposium on Ice Problems*, Helsinki, Finland, pp. 19–23.

Makarov, O., Gogoladze, D., and Bekker, A., 2023. Application of the Brazilian test method for testing the tensile strength of first-year ice in Novik Bay. *Ocean Engineering*, 269, 113488.

Malmgren, F., 1927. On the properties of sea ice. In Scientific results of the Norwegian north pole expedition with the Maud, 1918–1925, 1a(5). Bergen: a/s J. Griegs boktrykkeri, 1927, 67 p. Thesis, Upsala.

Petrich, C., and Eicken, H., 2016. *Sea Ice: Overview of sea ice growth and properties*. United Kingdom: John Wiley & Sons.

Ryvlin, A. Y., 1974. Method for predicting the ultimate flexural strength of an ice sheet. *Arctic and Antarctic Research*, 45, 79–86.

Schulson, E. M., and Duval, P., 2009. *Creep and fracture of ice*. Cambridge University Press.

Schulson, E. M., Fortt, A. L., Iliescu, D., and Renshaw, C. E., 2006. Failure envelope of first-year Arctic sea ice: The role of friction in compressive fracture. *Journal of Geophysical Research: Oceans*, 111, C11S25.

Tarovik, O., Yakimov, V., Dobrodeev, A., and Li, F., 2022. Influence of seasonal and regional variation of ice properties on ship performance in the Arctic. *Ocean Engineering*, 257, 111563.

Taylor, R., Turnbull, I., Bailey-Dudley, E., and Pritchett, R., 2021. Full-scale in-situ four-point beam bending field tests on sea ice. In *International Conference on Offshore Mechanics and Arctic Engineering*, Vol. 85178, p. V007T07A017. American Society of Mechanical Engineers.

Thomas, D. N., editor, 2017. *Sea ice*. John Wiley & Sons.

Timco, G. W., and Frederking, R. M. W., 1996. A review of sea ice density. *Cold Regions Science and Technology*, 24(1), 1–6.

Timco, G. W., and Weeks, W. F., 2010. A review of the engineering properties of sea ice. *Cold Regions Science and Technology*, 60(2), 107–129.

Wang, Q., Liu, Y., Lu, P., and Li, Z., 2023. The porosity effect on the mechanical properties of summer sea ice in the Arctic. *The Cryosphere Discussions*, 2023, 1–21.

Weeks, W. F., and Wettlaufer, J. S., 1996. Crystal orientations in floating ice sheets. In *The Johannes Weertman Symposium*, pp. 337–350. Warrendale, PA: The Minerals, Metals and Materials Society.

World Meteorological Organization (WMO), 2014. *Sea-ice nomenclature*, WMO/OMM/BMO Report 259. Geneva, Switzerland: World Meteorological Organization.

Yen, Y. C., 1981. *Review of thermal properties of snow, ice, and sea ice* (Vol. 81, No. 10). US Army, Corps of Engineers, Cold Regions Research and Engineering Laboratory.

3 Ship-Ice Interaction

3.1 PHENOMENA AND ENERGY FLOW

3.1.1 Phenomena at Different Scales

Ship-ice interaction involves phenomena at different scales, here named as the material scale, the local scale and the ship scale, see Figure 3.1. Ship-ice interaction takes place at the local scale, which attributes to the phenomena at material scale, and eventually determines the phenomena at ship scale.

At the material scale (in the order of 10^{-4}~10^{-2} m), the main issue of concern is the ice structure and material property. As a material formed by natural freezing, ice is composed of grains of various size, typically in the order of 10^{-2} m on the top layer of sea ice. The grains are columnar due to the downwards growth of sea ice. Therefore, the mechanical property of sea ice is typically transverse isotropic and changes across the thickness. Meanwhile, ice is not homogeneous everywhere. It contains a large number of flaws including air pockets and brine drainage channels. These flaws act as stress concentrators when ice is under loading, and excites the crack to propagate. The main phenomenon at the material scale in the context of ship-ice interaction is then the onset of cracks, which is affected by the physical parameters including temperature, salinity, brine volume and porosity.

At the local scale (in the order of 10^{-1}~10^{0} m), ice exhibits different failure modes, including crushing, flexure, shearing and splitting, which will be elaborated in the later sections of this chapter. These are macroscopic phenomena as the result of the crack propagation. The failure of ice with various modes converts the state of ice from continuum to discrete elements. Meanwhile, on the ship side, the structure deforms under the contact force and its structure members may yield or buckle if the corresponding limits are exceeded.

At the ship scale (in the order of 10^{1}~10^{2} m), the phenomena of concern are the dynamics of the ship and ice during the interaction. The summation of ice loads along the ship hull determines the global response of the ship and movement of the ice floe. Therefore, ship-ice interaction is a cross-scale problem. Typically, for such problems, proper assumptions and idealizations at smaller scales need to be made, in order to enable solutions at larger scales.

3.1.2 Energy Flow during Interaction

When a ship navigates through ice-covered water, the hull and propeller of the ship constantly interact with ice. The interaction incurs responses of both the ship and the ice, consuming a part of the kinetic energy of the ship through multiple means. The dissipation of energy determines the force of interaction, resulting in ice loads onto the structure and eventually affecting the structural safety and ship performance. Figure 3.2 illustrates the energy flow during this process.

DOI: 10.1201/9781003676829-3

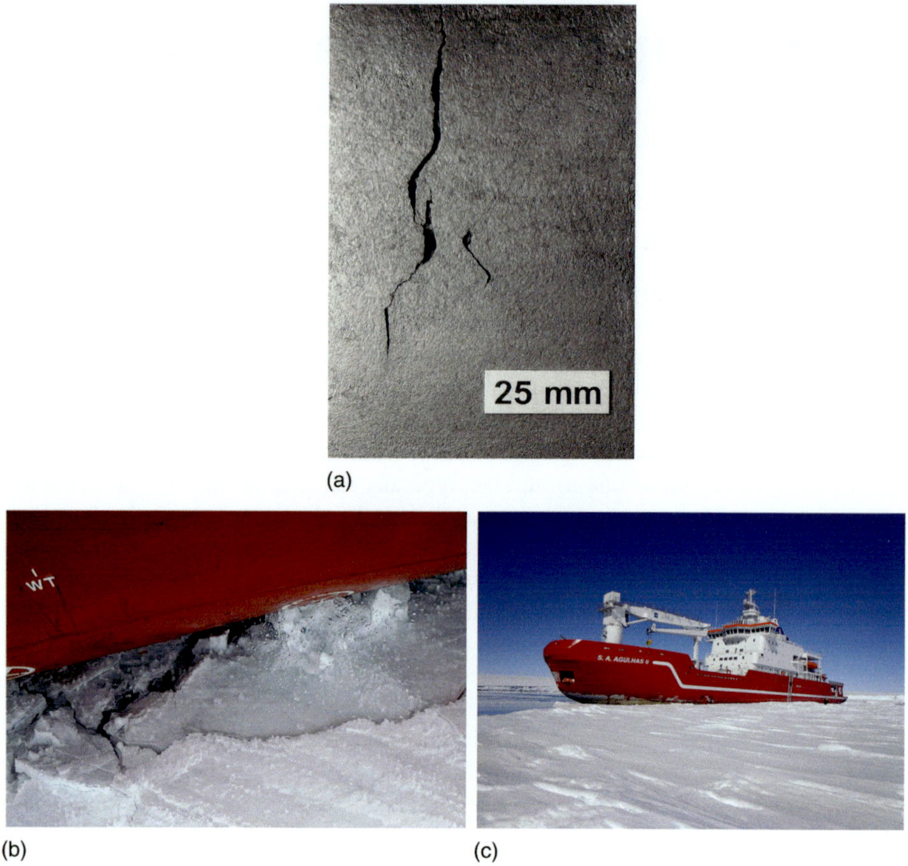

FIGURE 3.1 Phenomena at different scales related to ship-ice interaction, (a) crack onset at material scale (https://icelab.engineering.dartmouth.edu/media/photos); (b) bending failure at local scale; (c) ramming-induced trimming at ship scale.

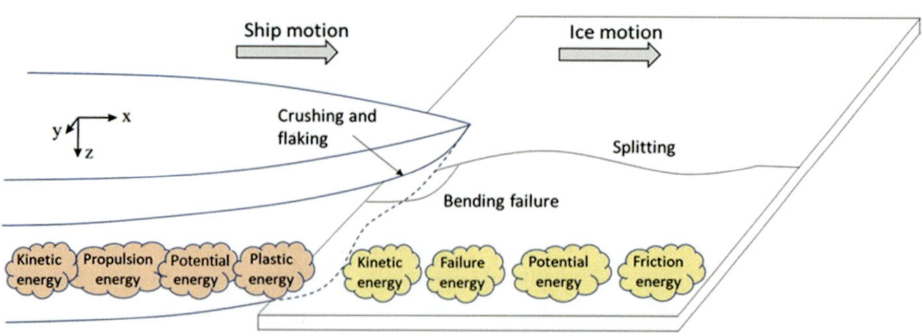

FIGURE 3.2 The energy flow during ship-ice interaction process, along with possible ice failure modes.

From the perspective of ice, under the contact force, ice exhibits different failure modes including crushing, flexure, shearing and splitting, as will be elaborated in later sections. As the result of its brittle nature, the failure of ice happens mainly through the formation and propagation of cracks of various size and direction. Meanwhile, the ice sheet deforms and the broken ice submerges under the contact force, which are restorable and therefore together classified as potential energy. In addition, the ship may push the ice aside and increase the kinetic energy of ice. Therefore, we can write the following equation denoting the energy consumed through ice:

$$E_{\text{ice}} = E^I_{\text{failure}} + E^I_{\text{potential}} + \Delta E^I_{\text{kinetic}} \tag{3.1}$$

where E_{ice} is the total energy consumed by ice, while E^I_{failure}, $E^I_{\text{potential}}$ and $\Delta E^I_{\text{kinetic}}$ are the energy consumed by ice failure, deformation and submersion, and motion, respectively.

The portion of energy consumed through different sources varies depending on the mass of ice floe, illustrated schematically in Figure 3.3. When the ice mass is small, ice floe has more mobility so the energy is mainly converted into kinetic energy. On the contrary, when the ice mass is very large (typically in a higher order of magnitude than the ship), the ice floe is stationary and the energy is mainly consumed through breaking and deflecting the ice sheet.

From the perspective of the ship, the structure deforms under the contact force, thereby gaining elastic strain energy, which is restorable and therefore can be seen as a type of potential energy. If the contact force exceeds the elastic limit of the structure, the structure undergoes plastic deformation and consumes energy through plastic work. Meanwhile, the propulsion system tends to increase the kinetic energy of the ship while the interaction process consumes it. In addition, the friction between ship and ice converts a part of the energy into heat. The following equation of energy conservation can be established:

$$E^S_{\text{propulsion}} + \Delta E^S_{\text{kinetic}} = E^S_{\text{potential}} + E^S_{\text{plastic}} + E^I_{\text{potential}} + E^I_{\text{failure}} + \Delta E^I_{\text{kinetic}} + E_{\text{friction}} \tag{3.2}$$

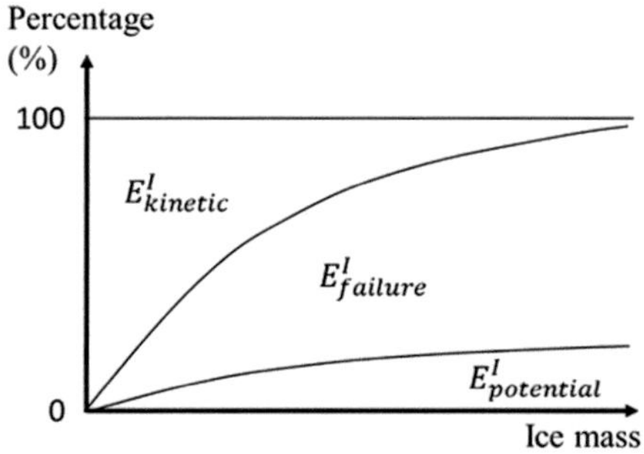

FIGURE 3.3 Source of energy consumed through multiple means.

where $E^S_{\text{propulsion}}$ denotes the energy gained from the propulsion system during the interaction process; $\Delta E^S_{\text{kinetic}}$ is the kinetic energy change of the ship before and after the interaction; $E^S_{\text{potential}}$ is the potential energy containing both the elastic strain energy and the potential energy due to heave and pitch; E^S_{plastic} is the plastic energy of the hull; E_{friction} is the energy consumed through friction between the ship and ice.

The portion of energy consumed by the ship and ice varies depending on the initial kinetic energy of the ship-ice system. Let us consider a ship interacting with a large ice floe. Here "thick" means crushing is the only failure mode (no bending failure) and "large" means the ice is stationary during the interaction ($\Delta E^I_{\text{kinetic}} = 0$). Following Newton's third law, the force acting on the ship and on the ice are the same in magnitude and opposite in direction. Under the contact, the structure deforms while the ice gets crushed, as illustrated in Figure 3.4. The contact force as functions of the structural deformation and crushing indentation can be plotted schematically in Figure 3.5. The energy consumed during the interaction process is then the summation of the shaded area below the two curves (here neglecting $E^I_{\text{potential}}$ which is much smaller than E^I_{failure}). When the initial kinetic energy is small, the structure is in elastic stage and the energy is mostly consumed by ice. The structure can be regarded as a rigid wall with no deflection. However, when the initial kinetic energy is large, the structure undergoes plastic deformation, which then becomes one major source of energy consumer.

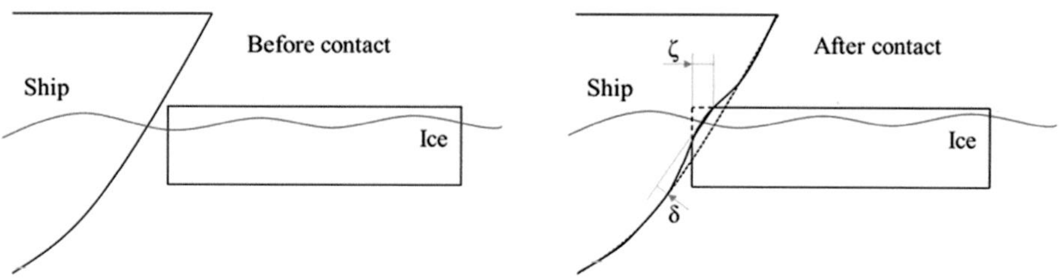

FIGURE 3.4 Sketch of a ship interacting with a thick, large ice floe, with structural deformation δ and crushing indentation ζ.

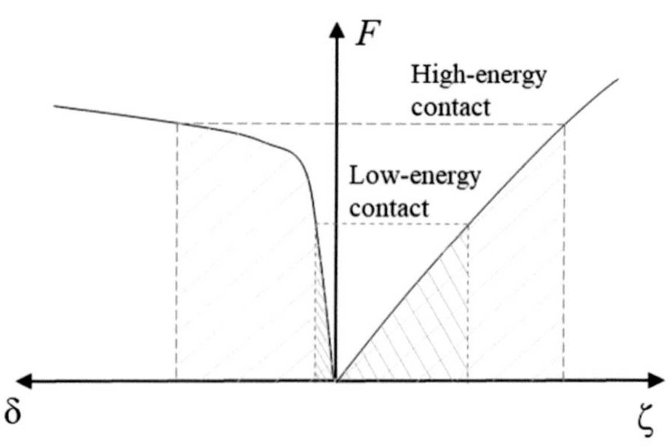

FIGURE 3.5 Energy absorbed by ship through deformation and by ice through crushing, under low- and high-energy contacts, the left x-axis is the ship deformation and right one the ice indentation depth.

3.2 ICE FAILURE AND MOTION

The process of ice failure and movement plays a major role in the limiting mechanism of ice loads. In this section, we will go through the main failure modes of sea ice as well as the movement of discrete ice after failure.

3.2.1 Crushing, Flaking and Contact

The uniaxial strength of ice is in the order of 10^0 MPa. In comparison, the yield strength of steel is in the order of 10^2 MPa. As a result, the contact between ships and ice is exerted by breaking the ice edge. This is essentially different from the situation of elastic impact such as two metal balls colliding with each other, generating a force pulse within very short duration (~10^{-2} s). In ship-ice contact, the duration of each contact scenario is in the order of 10^{-1}–10^0 s. The contact force is then determined mainly by the strength of ice, rather than the elasticity of the contact bodies.

3.2.1.1 Crushing and Flaking

Crushing is the most important failure mode of sea ice, which is prevalent in almost all ice-structure interaction scenario. The process of crushing determines the pressure on the contact area and hence the contact force. Figure 3.6 shows the contact between a ship and ice, from which we can see powder-like ice particles ejecting from the contact region, as the result of ice crushing.

Why does crushing happen? As introduced in Chapter 2, ice behaves as a brittle material under ship-ice interaction, therefore is prone to fail by fracture rather than yielding. When loaded by a ship at its edge, numerous cracks are formed in the contact region, breaking ice into small particles. Some of the particles are ejected from the contact region, while the rest are trapped in between the ship and the intact ice,

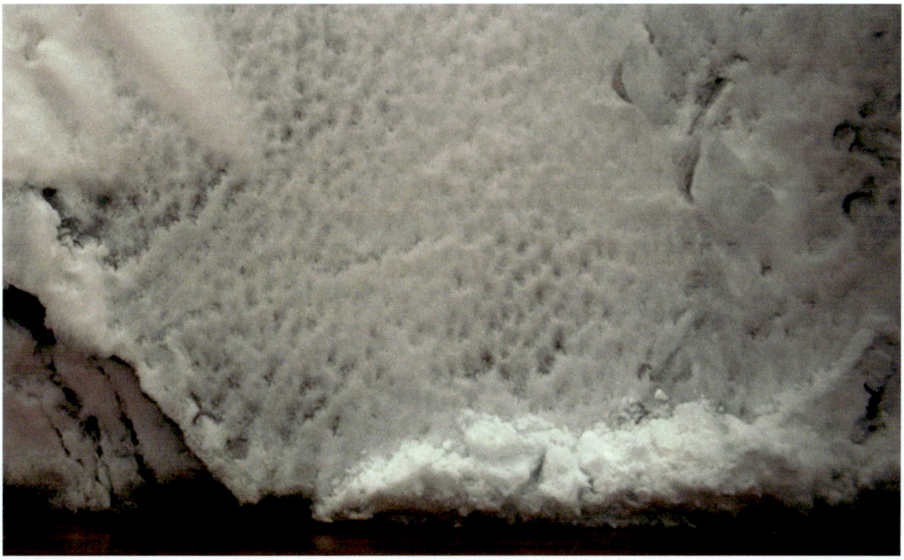

FIGURE 3.6 Crushed ice ejecting from the contact region by a ship.

as illustrated in Figure 3.7. The contact force between the ship and ice is then transmitted via two means: direct contacts and crushed ice. The direct contacts are discretely located on the contact area, which forms high-pressure zones (HPZs) that carry most of the contact force.

Crushing is not the only failure mode which occurs on the contact area. Another important process is flaking. When ice is loaded as illustrated in Figure 3.8, the contact force incurs shear stress in the direction of loading. The shear stress aggregates the microcracks along the shear direction and forms shear fracture, eventually breaking a small piece of ice which drops off from the intact ice sheet. This process is called flaking. Flaking results in a sudden drop of the contact force due to the reduction in contact area. The time history of the contact force can then behave like saw-tooth.

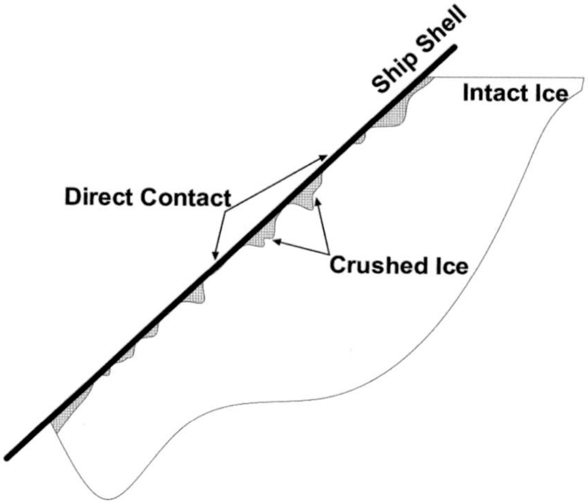

FIGURE 3.7 Ship-ice interaction through direct contact and crushed ice.

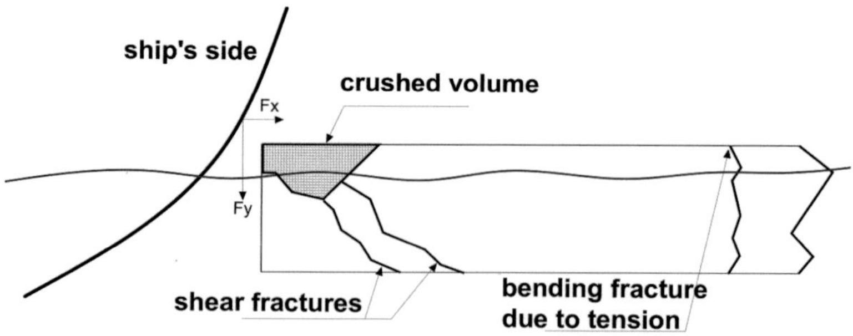

FIGURE 3.8 Ice edge failure process as described in Varsta and Riska (1977), where flaking of ice pieces is formed by shear fracture.

(Re-produced from the original.)

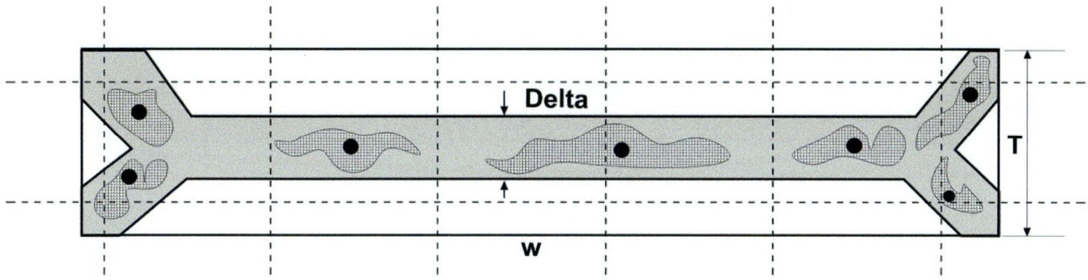

FIGURE 3.9 Line-like feature of the contact area and distribution of the HPZs on the nominal contact area.

(Re-produced from Taylor et al., 2019.)

For cylinder structures in ice such as wind turbine, modelling of the flaking process is crucial because the periodic force fluctuation can incur resonant of the structure, when the frequency of flaking is close to the natural frequency of the structure. For ships sailing in ice, the process of flaking is usually not explicitly modelled, and instead is implicitly considered in the contact pressure calculation. However, flaking has direct effect on the contact area. In relatively thin ice, it has been observed that the contact between ship and ice is line-like, as illustrated in Figure 3.9 by the shaded area. This is because the flaking of small ice pieces reduces the contact height and thereby the contact area. In Finnish-Swedish Ice-Class Rules (FSICR), the contact height (i.e., the *delta* in Figure 3.9) is defined as 30% of the prevailing ice thickness (i.e., the *T* in Figure 3.9), to account for the line-like feature (see also Figure 6.5).

3.2.1.2 Contact Area and Pressure

In the context of ship-ice interaction, the main interest is to know what is the pressure transmitted between the ship and ice and what is the size of the contact area. However, as described above, ice crushing and flaking are rather complex processes. It is therefore very difficult to estimate the actual contact area and pressure.

There are three different types of contact area, which must be distinguished to understand and estimate the contact pressure. These are termed as the nominal contact area, the true contact area, and the design area, illustrated in Figure 3.10 and explained below:

- Nominal contact area: The area of overlap between the ice and structure, which is usually calculated from the indentation depth and the known geometry of the edge. The nominal area is a concept to enable pressure estimation, especially for the calculation of global ice loads, but is not the true contact area as introduced below.
- True contact area: The area of actual contact through direct contact and crushed ice, which is the summation of the discrete contact spots. This accounts for the ice failure including crushing and flaking.

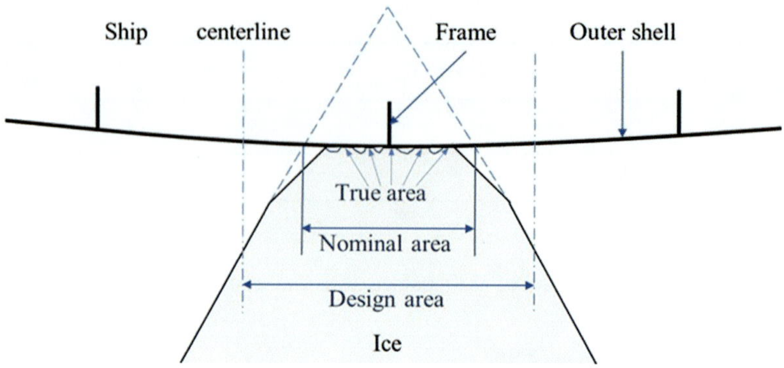

FIGURE 3.10 The nominal contact area, true contact area and design area, as well as the corresponding pressure.

- Design area: A fixed area of interest, which is defined according to the layout of the structural members, to be used for structural design. For ships, this can be a rectangular load patch with length of a frame spacing and a defined height. The size of design area is independent of the contact, and can cover or be a part of the true or nominal contact area.

In the history of ice mechanics research, many attempts have been done to model ship-ice contact and estimate the pressure. The most widely adopted approach to date, however, is still empirical approach. In most of the ice-class rules or standards, the reference pressure on a design area is given as constants obtained through full-scale measurements, and is then modified based on the geometry of the design area, ship speed or engine power. For example, Finnish-Swedish Ice-Class Rules (FSICR) define the design pressure by Eq. (3.3)

$$p = c_d c_p c_a p_0 \tag{3.3}$$

where p_0 is a reference pressure being 5.6 MPa; c_d is the coefficient accounting for the size and engine output of the ship; c_p is for the magnitude of the load expected in the hull area in question relative to the bow area; c_a accounts for load width, see also Chapter 6, Section 6.2.1. There exists a pressure-area relationship, stating that the larger a design area is, the smaller the average pressure will be. This is reflected in the parameter c_a, which is calculated as per on Eq. (3.4)

$$c_a = \sqrt{\frac{l_0}{l_a}}, \text{ and } 0.35 \le c_a \le 1 \tag{3.4}$$

where l_0 is a reference width being 0.6 m; l_a is the width of the design area. As can be inferred, the pressure in the design area, p, is proportional to $l_a^{-0.5}$, hence decreases with the design area.

Why the pressure-area relationship exists? The reason can be multiple, and debatable. Here we do not attempt to elaborate all the possible reasons of that, but state one simple reason. The contact width of ship-ice contact is high random and variable. Consider a design area of small load width, there is a higher probability that the whole area is inside the actual contact area, while with longer load width, the probability of being fully loaded is smaller. This results in higher chance of larger pressure to be measured on smaller areas, which can lead to the well-known pressure-area relationship. Figure 3.11 shows a compilation of measured pressure on local design areas as per Masterson et al. (2007). Readers who are interested in this topic are referred to the literature, such as Schulson and Duval (2009), Daley (2007), Timco and Sudom (2013) and Palmer et al. (2009).

For a nominal contact area, opinions on its relationship with pressure vary among the literature. Some support a slight decrease in pressure when nominal contact area increases, while some support a constant or slight increase (Daley, 2007). One simple way of estimating the contact pressure is to link it to the uniaxial compressive strength, by assuming the contact pressure equals the uniaxial compressive strength by a coefficient, i.e., Eq. (3.5).

$$p = c\sigma_c \tag{3.5}$$

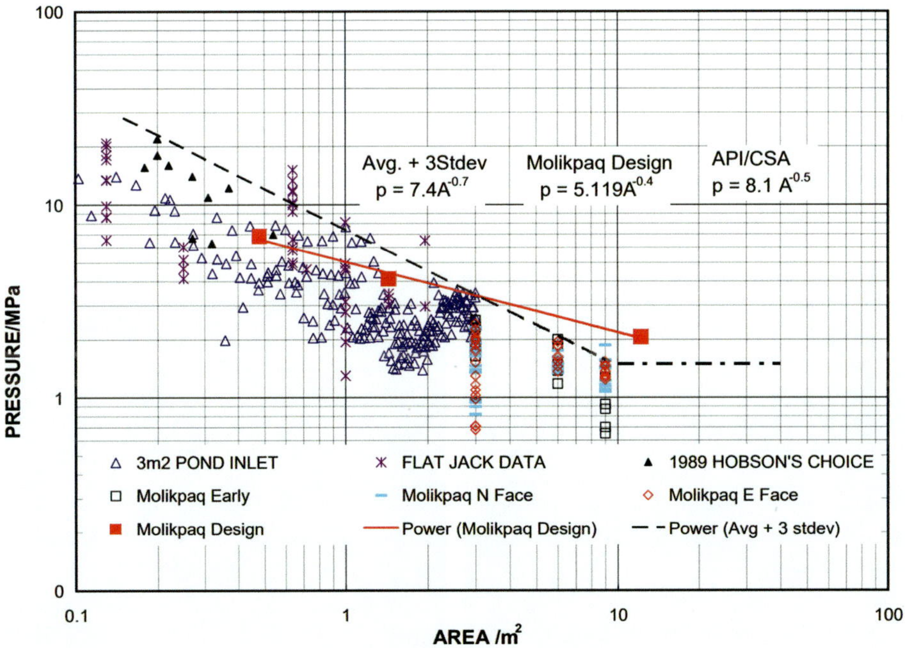

FIGURE 3.11 Pressure-area relationship, as per Masterson et al. (2007), scatters showing measurement results from different sources.

where σ_c is the uniaxial compressive strength and c is the coefficient. ISO 19906 (ISO, 2019) suggests Eq. (3.6), to calculate nominal contact pressure for vertical structures

$$p = C_R \left(\frac{h}{h^*}\right)^n \left(\frac{w}{h}\right)^m \quad (3.6)$$

where C_R is an ice strength coefficient which can relate to the uniaxial strength, w is the structure width and h the ice thickness; h^* is a reference thickness; n and m are coefficients with negative values, which correspond to decreased pressure as a function of nominal contact area.

Although there is indeed correlation between the contact pressure and compressive strength, the loading state of ice is apparently different between ship-ice contact and uniaxial compression. In the former case, the surrounding ice exerts high confinement and hence the ice is under triaxial loading, resulting in higher load than the uniaxial case.

The estimation of contact pressure distribution on the true area, however, is very challenging and has not reached consensus. It is well known that the loads are exerted through discrete high-pressure zones. But modelling the occurrence and disappearance of the high-pressure zones require extensive efforts. Pressure on the nominal area and design area are more commonly used in practice.

3.2.2 Bending Failure and Splitting

When loaded by the ship, a stress field is activated as the result of the contact force. Due to the vertical (out-of-plane) component of the contact force, the distribution of stress at the top and bottom surfaces of the ice floe is different. Figure 3.12 presents the distribution of maximum principal stress from a Finite Element Analysis. On the top surface, the maximum principal stress in tension tends to open cracks in the circumferential direction, while on the bottom surface it tends to open in the radial

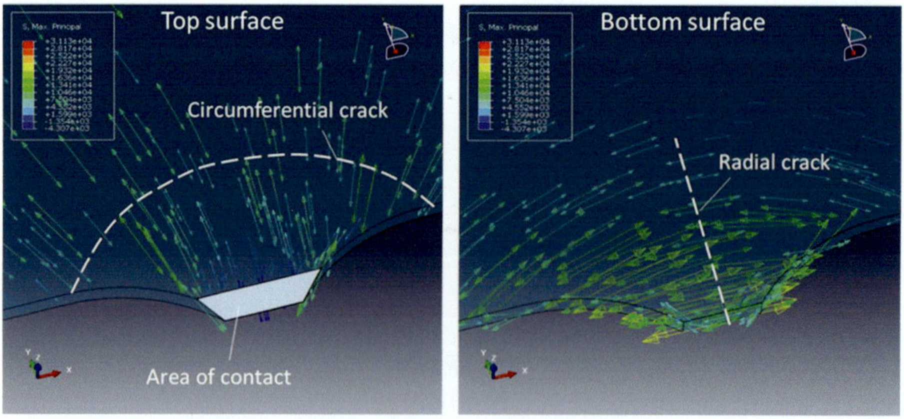

FIGURE 3.12 Circumferential crack due to bending, initiated at the top surface (left), and radial crack due to splitting, initiated at the bottom surface (right); the arrows denote the direction of maximum principal stress as the result of the contact force.

direction. This results in two different failure modes, namely bending failure and splitting. These two failure modes compete, and one of them dominates depending on various parameters including at least ship speed, ice geometry and anisotropy in mechanical properties.

3.2.2.1 Bending Failure

Ice floe of relatively large size is prone to fail by bending. The inclination of ship hull incurs a downward force component, which bends the ice sheet downwards and results in tensile stress above the neutral axis and compressive stress below. As a brittle material, ice fails more easily under tension. Cracks are then prone to occur at the location where the tensile stress is maximum, which is located at the top surface. As illustrated in Figure 3.12, the crack will open and propagate in the circumferential direction, throughout the surface and meanwhile extending through the thickness. This results in an ice piece dropped off its parent ice sheet, which continues to interact with the ship as a discrete object. Figure 3.13(a) shows a photo of a ship model breaking level ice at Aalto Ice Tank. As can be seen, multiple circumferential cracks take place around the ship body, which results in discrete ice pieces broken off from the intact ice sheet, and leaves space for the ship to move ahead. The occurrence of circumferential cracks is the key events which converts the intact ice sheet into discrete ice pieces, and determines the level of ice loads and icebreaking resistance.

It can be naturally concluded that tensile strength is the key mechanical property determining whether bending failure happens. As stated in Chapter 2, Section 2.4, the tensile strength of sea ice is rather difficult to measure. More commonly, the flexural strength is measured to approximate the tensile strength, by cantilever beam test, three-point bending or four-point bending test. Flexural strength is also the key parameter in ice model testing to maintain the scalability to full scale.

For the estimation of ice loads and resistance, there are two important parameters which need to be reasonably calculated, termed as the bearing capacity and the ice piece size. Bearing capacity is defined as the maximum vertical force ice can bear when bending acts as the dominant failure mode. Conventionally, this term has been established to describe the maximum load such as a truck or a helicopter on an ice sheet. As sketched in Figure 3.13, in the context of ship-ice interaction, the difference is that the load is exerted through an inclined structure. This generates an in-plane compressive force in addition to the vertical force, which can have major influence on the bearing capacity.

Ice piece size, as indicated by its name, is the size of ice piece formed through circumferential cracks. This is often characterized by the distance from the contact area to some position on the circumferential crack. An example of the ice piece measurement is given by Figure 3.14.

The bearing capacity and ice piece size are affected by multiple factors, listed below:

- Flexural strength. The bearing capacity is often assumed to be proportional to the flexural strength under quasi-static loading. The ice piece size is positively correlated to flexural strength.
- Ice thickness. The bearing capacity is often assumed to be proportional to the square of ice thickness. The ice piece size is positively correlated to ice thickness.

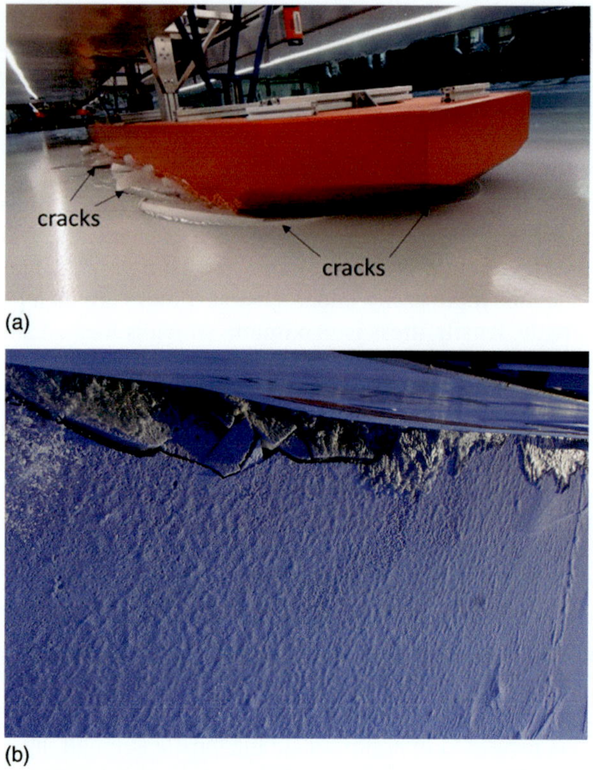

FIGURE 3.13 Circumferential cracks (a) generated by bending failure around the ship body in a model scale test, and (b) observed in full-scale test.

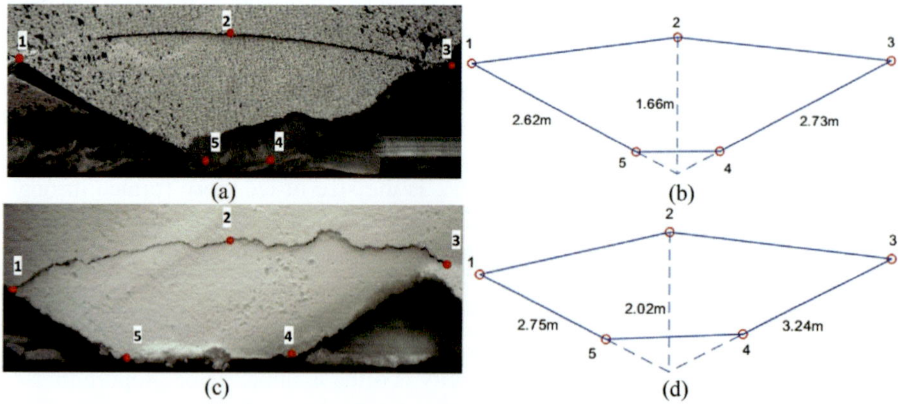

FIGURE 3.14 Measurement of the size of ice pieces. The numbers denote the distance between point 1, 2 and 3 to the virtual tips obtained by prolonging and intersecting line 1–5 and 3–4 (Li et al., 2019).

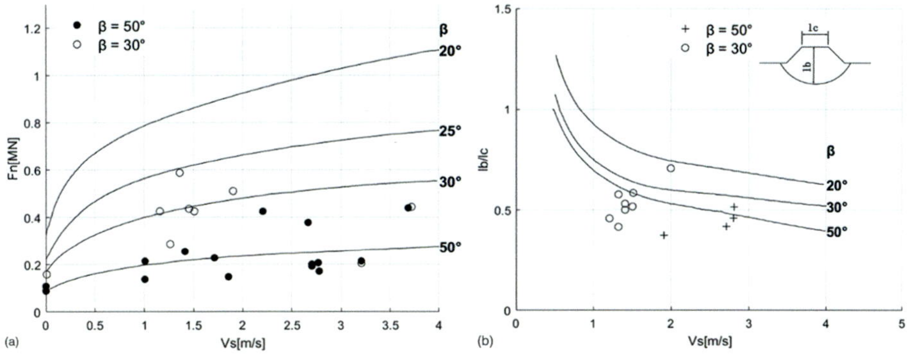

FIGURE 3.15 The bearing capacity as a function of loading speed with different frame angles (a), and the icebreaking length as a function of loading speed with different frame angles (b). The scatter denotes data obtained from in-situ measurements, and the curves denote Finite Element Analysis results by Varsta (1983).

(Re-produced from the original.)

- Elastic modulus. The bearing capacity is negatively correlated to elastic modulus. The ice piece size is positively correlated to elastic modulus.
- The inclination angle. Larger inclination angle incurs higher in-plane compressive force (F_x in Figure 3.15) in comparison to vertical force (F_y in Figure 3.15). As a result, both the bearing capacity and ice piece size increase. The influence is small at low speed and more evident at high speed, as indicated by Figure 3.15.
- The loading speed. It is well known that there exists a dynamic effect in the ice sheet as the result of rapid loading. This magnifies the bearing capacity but reduces the ice piece size. Evidences of such effect are shown in Figure 3.15.

The ice loads and resistance due to bending failure are then dependent on the above parameters via bearing capacity and ice piece size.

3.2.2.2 Splitting

As shown in Figure 3.16, when ships navigate in ice floe fields, in addition to circumferential cracks caused by bending fracture, ice floes often fail by radial splitting under the action of contact forces. An example of this is given in Figure 3.17. Compared with circumferential cracks, the propagation of radial cracks is not limited to the vicinity of the contact area but extends to other free boundaries under the continuous action of contact forces. Their length is much larger than that of circumferential cracks, and a large amount of energy is consumed in the crack propagation process. Therefore, it is not possible to determine the occurrence of radial cracks based only on the initial conditions of the cracks. It is necessary to use fracture mechanics which can estimate the crack evolvement.

Comparing to bending failure, splitting has received much less attention in the history of research on ice-going ships. This is because bending failure is dominant

FIGURE 3.16 Radial crack induced by splitting, top-left: before crack initiation, top-right: after crack propagation, bottom: floe split and separated.

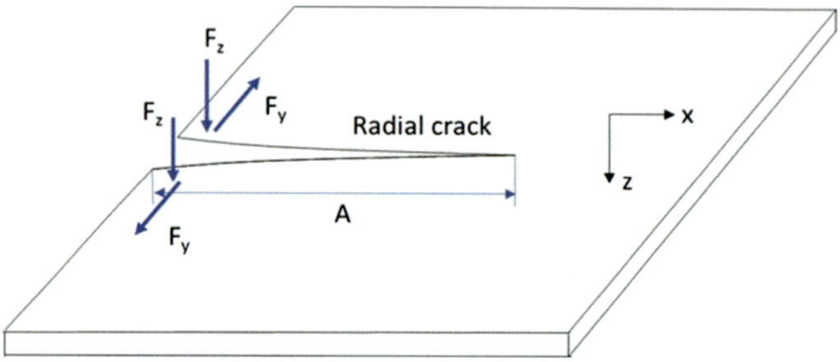

FIGURE 3.17 Splitting incurred by in-plane and out-of-plane force.

when ships sail in large ice floe (size in the order of 10^3 m or more), which is usually used as the design ice condition. However, merchant ships are nowadays increasingly optimized for open water as required by the environmental regulations to reduce carbon emission, which compromises the ice-going performance. The performance in ice floe fields determines more of the icebreaking performance for such ships. It is therefore important to understand how ice fails in splitting.

To date, there has not been a well-established method to calculate the evolvement of radial cracks. Therefore, here we merely present the basic theory behind this phenomenon via linear elastic fracture mechanics (LEFM), following the theoretical investigation by Lu (2014).

Assume a radial crack of length A has been created as the result of ship-ice contact, as shown in Figure 3.17. According to LEFM, to further extend the crack, the energy release rate, $G(A)$, should reach the critical energy release rate, G_{IC}, i.e. Eq. (3.7)

$$G(A) = G_{IC} \tag{3.7}$$

The critical energy release rate, G_{IC}, is a material parameter. The energy release rate, $G(A)$, denotes the energy released by further opening the crack by a unit length/area. According to Irwin's theory, $G(A)$ is a function of the strain energy of the ice floe, denoted by U, and external energy, denoted by T, by Eq. (3.8)

$$G(A) = -\frac{d(U - T)}{hdA} \tag{3.8}$$

This is the equation governing the crack propagation process. The key issue is then to solve $\frac{dU}{dA}$ and $\frac{dT}{dA}$, i.e., the change of strain energy and external energy per unit extension of crack length. These can be solved either theoretically or numerically. Here we do not further open this and refer to Lu (2014) for readers who are interested in more details of possible solutions.

3.2.2.3 Competition between Bending Failure and Splitting

As mentioned in the beginning of this subsection, there exists competition between bending failure and splitting. Under different conditions, either exhibits as the dominant failure mode. The determination of failure modes depends on multiple factors, including the following:

- Size of the ice floe. Radial cracks occur mostly in medium-sized ice floe. This is because radial cracks tend to propagate towards other free edges, where there is the least confinement. In large-sized ice floe, the confinement is so high that the propagation of radial cracks is restricted. In small-sized ice floe, ice simply moves away without radial or circumferential cracks. In addition to the volume of ice floe, the actual shape also affects the failure mode.
- Ship speed. Higher interaction speed tends to increase the ratio between the magnitude of tensile stress on the top surface and that at the bottom surface. This results in higher possibility of bending failure at high speed, and higher possibility of splitting at low speed.
- Anisotropy of material property. The mechanical property of ice is not constant across the thickness. The actual distribution depends on the temperature and salinity across the thickness. This affects the stress on the top and at the bottom surface, influencing the actual failure mode.

The actual failure mode is then the result of factors including at least the above. In general, when ships navigate in level ice field of very large ice floe (~10^3 m), bending failure is of primary concern. In medium-sized ice floe (~10^1 m), splitting is prevalent. In between, both should be considered. The readers are referred to Lu (2014), Xu et al. (2020) and Kim et al. (2022) for more investigation on the competition between bending failure and splitting.

3.2.3 Buckling and Rubble Accumulation

Crushing, flaking, bending, and splitting are the main ice failure modes which take place when a ship navigates through ice fields of various floe size. Through these processes, large ice floes are broken into smaller floes or pieces which can be pushed away by the ship, making space for the ship to advance. Due to the inclined ship hull at the bow, ice rarely fail by buckling when the ship is going forward.

However, there is one special situation in which the ship extensively contacts ice at its midship section, namely the dynamic compressive ice field. Figure 3.18 shows a sketch of such scenario, where the ice drifts from the lateral direction. In such scenario, the midship part where the hull is vertical or close to vertical gets

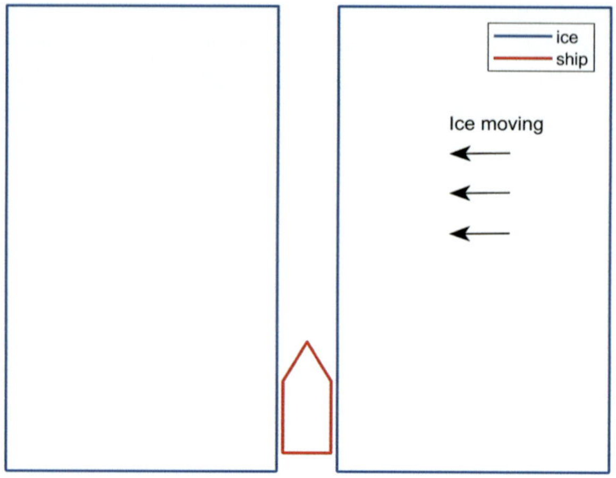

FIGURE 3.18 A ship in dynamic compressive ice field (Li et al., 2019).

FIGURE 3.19 Ice buckling (left) and rubble accumulation (right) at the midship, screenshot from a video record during SAFEWIN project model test.

heavily interacted with ice, which may break the ice through buckling. The ship may be able to advance, but often get stuck and cannot move.

Figure 3.19 shows a screenshot taken from a video record of a model test in the SAFEWIN project at Aalto Ice Tank. This may look similar to bending failure, which also breaks off an ice piece. The essential difference is that the maximum deflection of the ice sheet occurs at the location of crack in buckling, while in bending it occurs at the ship-ice contact location. In this case, the ship hull is vertical at the midship and cannot induce a vertical component of contact force sufficient to break ice by bending.

Another phenomenon, which happens when the ship is stuck in dynamic compressive ice fields, is the accumulation of ice rubble. This is caused by the continuous breaking of ice into pieces and accumulation against ship body, since the ship is stationary. Figure 3.19 shows such case in the SAFEWIN project model test. The ship is set stationary, and ice gets pushed towards the ship. It can be observed that a large number of broken ice pieces are accumulated. In real world, this can lead to significant additional resistance and make it extremely difficult for the ship to escape.

To date, comparing to other ice failure modes, buckling and rubble accumulation are not much investigated. Readers are referred to Kujala (1993) and Zhou (2012) for further investigations on these topics. In practice, buckling and rubble accumulation are typically not considered during ship design process.

3.3 STRUCTURAL RESPONSE

Section 3.2 introduced the response of ice during ship-ice interaction. From the perspective of ships, the structure members undergo extensive deformation. As discussed earlier in Figure 3.5, when the total energy of contact is low, the structure behaves elastically and absorbs little energy compared to ice. In such cases, the structure can be assumed as rigid for the calculation of ship-ice interaction. When the total energy of contact is high, the deformation reaches plastic stage, which can consume comparable amount of energy to that by ice. The structural deformation can then have major effect on the interaction process and must be considered in the modelling of ship-ice interaction.

Figure 3.20 shows the typical form of the local structure of icebreakers at the side of the bow. The structure is composed of frames in between webframes and stringers across the frames. When loaded by ice, the contact area enlarges gradually and can

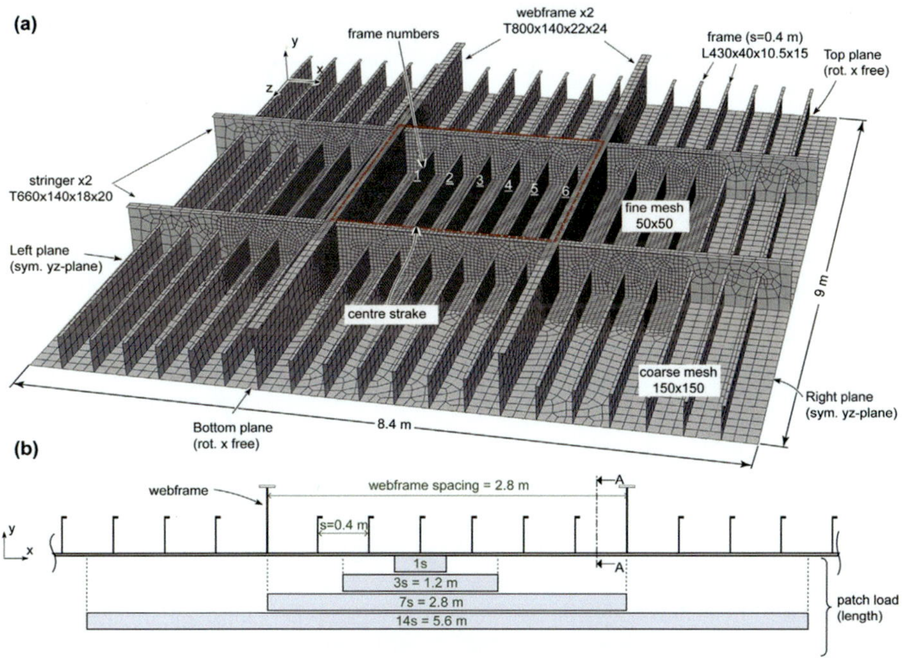

FIGURE 3.20 The typical form of the local structure of an icebreaker, and the evolving of ice loads on the structure (Kõrgesaar et al., 2018).

extend to cover several frames at the same time. Meanwhile, due to the forward motion of the ship, the area of ice-loading moves towards the aft of the ship, which also affects the response of ship structure (Quinton, 2019).

It has been observed that the ice pressure on a frame can be higher than on the shell plating at the midspacing between frames. This is because the flexural stiffness of frames is higher than shell plating. In addition, when one frame is loaded by ice, the force is passed to its adjacent frames through the shell structure. Therefore, the load is carried jointly by the frames in contact with ice and the adjacent frames, along with the shell. The actual load distribution is schematically shown in Figure 3.21 as per Suominen et al. (2017), which indicates that the region of structure response is larger than the nominal area of contact.

Different structure damage modes can take place under ice loading. The most typical one for ships with transverse frames is the so-called "thin horse" damage, where the outer shell gets more heavily indented than the frames. A photo of such damage is shown in Figure 3.22. Buckling and shearing often take place in ship frames in addition to plastic bending. Under excessive ice loads, fracture may occur at the outer shell.

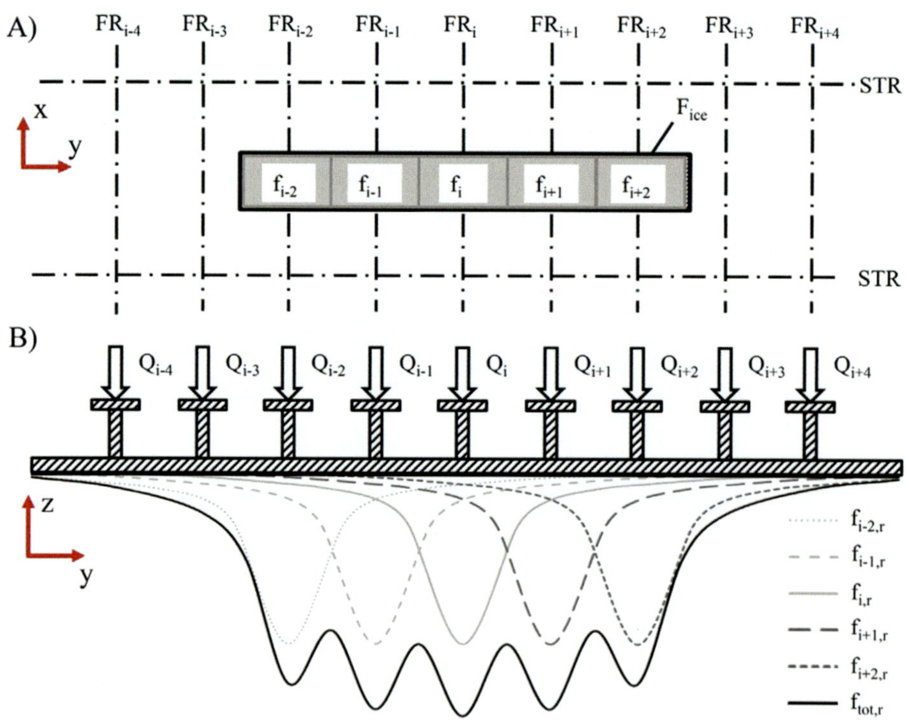

FIGURE 3.21 Load distribution on the structure during ship-ice interaction, (a) layout of the structure and region of ice pressure; (b) load carried by different frames and the connecting shell (Suominen et al. 2017).

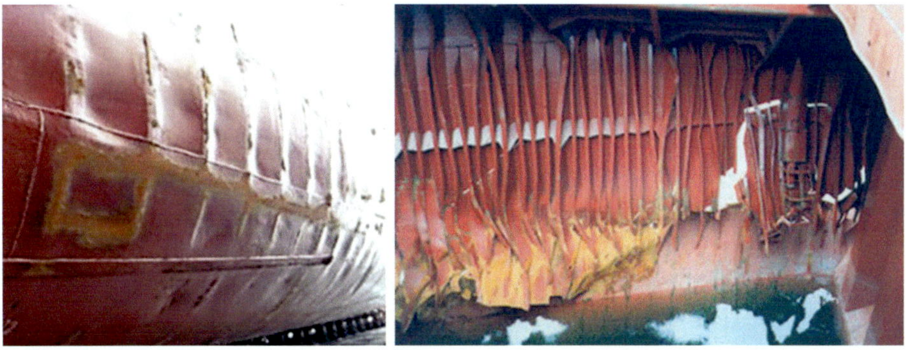

(a) (b)

FIGURE 3.22 Typical ice damages on ship structure. Left: dents on the outer shell; Right: buckling of frames (Hänninen, 2001).

The calculation of structural response is relatively straightforward with the Finite Element Method (FEM) if the ice pressure distribution is known. As a common practice specified by ice-class rules, a load patch of given shape, usually rectangular, is assumed and assigned with uniform pressure. The structural response can then be computed with FEM. Following this procedure, the stress and deformation of the structure can be with the permitted ones to assess the safety of the structure with given scantlings. Figure 3.23 gives an example of the plastic deformation of structure under ice loading using FEM by Kõrgesaar et al. (2018), with the load deformation curve schematically shown in Figure 3.24. As can be observed, ship structures typically have large overload capacity. The structure can still bear much higher load after its first yielding before fracture occurs. Therefore, the selection of structural limit states matters significantly in the design of ice-strengthening structures as also studied systematically by Valtonen et al. (2020). Different structural limit states correspond to different levels of structural damage, which links to the level of ice load and eventually ice condition.

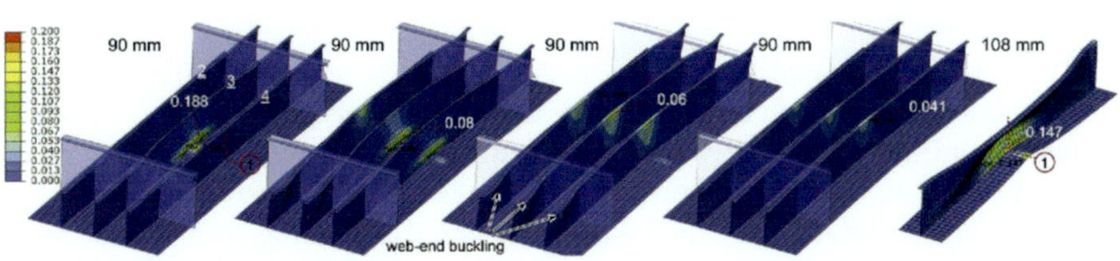

FIGURE 3.23 Plastic deformation of the structure under ice loading using FEM. The contours show the equivalent plastic strain. The four figures show simulation under different load width (Kõrgesaar et al., 2018).

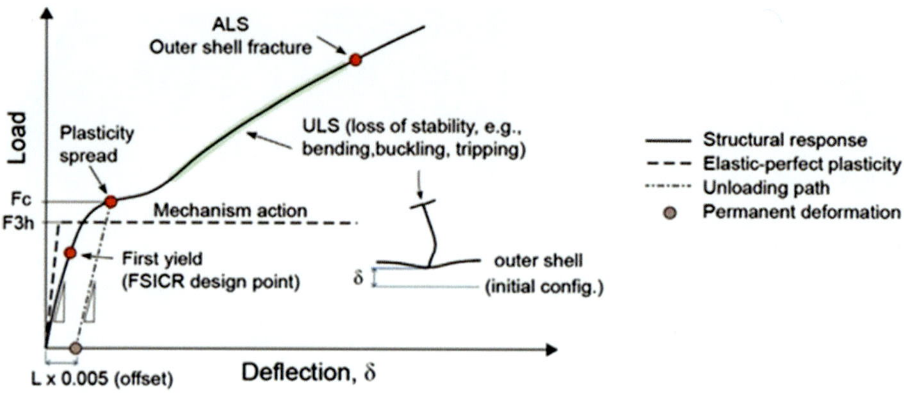

FIGURE 3.24 Example of load-deflection curves for frames. Permanent deformation refers to inelastic, plastic deformation that is not recoverable and is measured on the outer shell at the location of frame (Kõrgesaar et al., 2018).

3.4 PROPELLER-ICE INTERACTION

3.4.1 Historical Preview

When navigating in ice-covered waters, propellers are subjected to complex ice-water-propeller multi-phase coupling effects (see Figure 3.25). Their interaction with sea ice primarily occurs in two forms: indirect and direct interactions. When the sea ice is at a certain distance from the propeller, indirect interactions dominate, with sea ice mainly disturbing the flow field around the propeller. In the case of direct interaction, the propeller comes into direct contact with ice blocks, resulting in substantial ice loads on the propeller. This not only reduces the propulsion efficiency of the vessel but also poses a risk of severe damage to the propeller blades, thereby threatening the safety of the vessel. The direct interaction between the propeller and ice includes two forms: collision and milling. Of these, milling poses the greatest threat to the propeller. During the milling process, the propeller blades are in continuous contact with the ice and perform cutting operations. This process exerts significant cutting forces on the propeller blades, leading to severe surface wear and potential structural damage, which greatly affects the propulsion efficiency of the propeller and the safe operation of the ship. Comparing to the hull-ice interaction described in Sections 3.1 and 3.2, the impact speed between propeller and ice is remarkably higher causing different failure mechanism of the ice as described in the section.

Jussila and Soininen (1991) have prepared a comprehensive state-of-art report related to the propeller-ice interaction, and this is shortly reviewed next. Most of the existing propeller-ice interaction models are based on continuous loads due to the failure in a milling-type contact scenario (see Figure 3.26).

The first published propeller-ice interaction calculation method was the classical Jagodkin (1963) model. In his model, the first component is the load from the propeller face side due to the shear failure of ice. The other load component is caused by crushing failure generated by the blade projection in the pitch direction. Simple

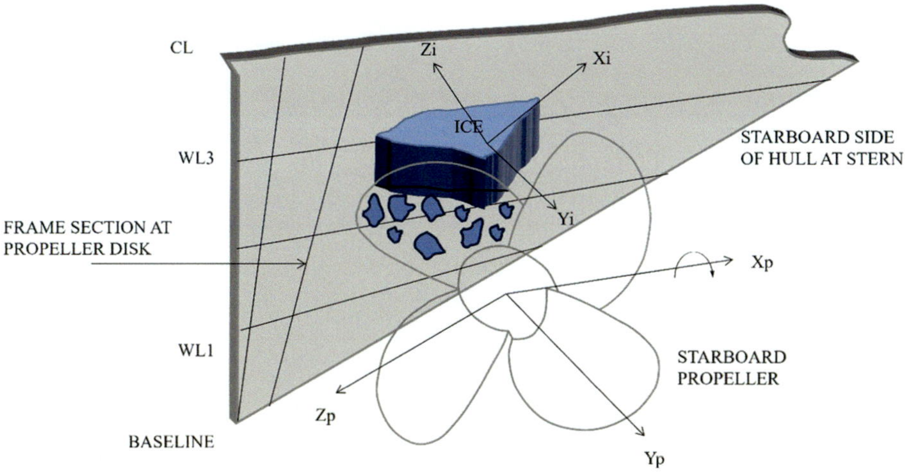

a) VIEW OF INTERACTION EVENT AT STARBOARD PROPELLER

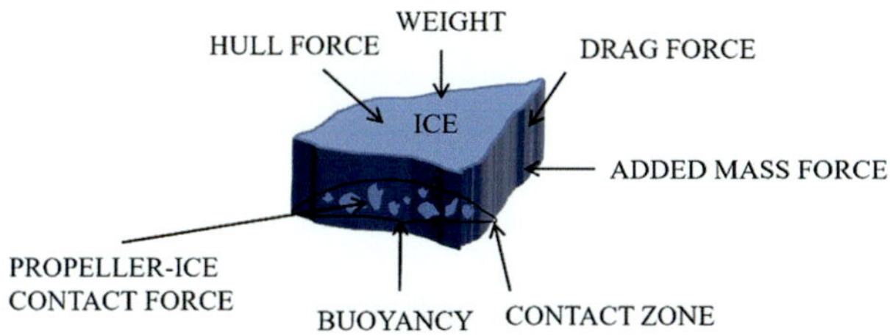

b) FREE BODY DIAGRAM OF ICE BLOCK

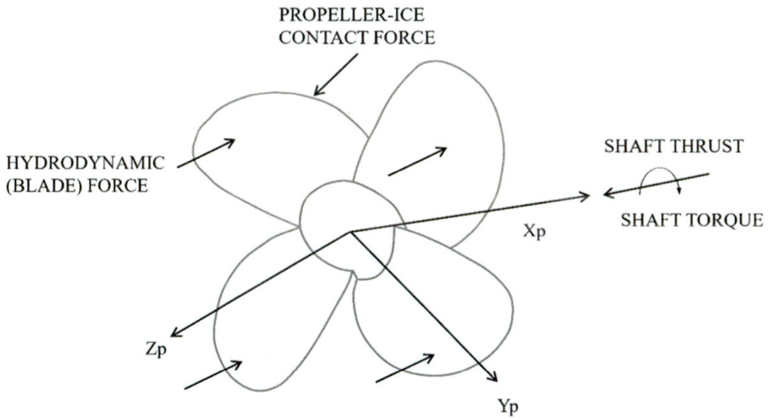

c) FREE BODY DIAGRAM OF PROPELLER

FIGURE 3.25 Affecting parameters on propeller-ice interaction. (a) View of interaction event at starboard propeller. (b) Free body diagram of ice block. (c) Free body diagram of propeller. (Re-produced from Veitch, 1995.)

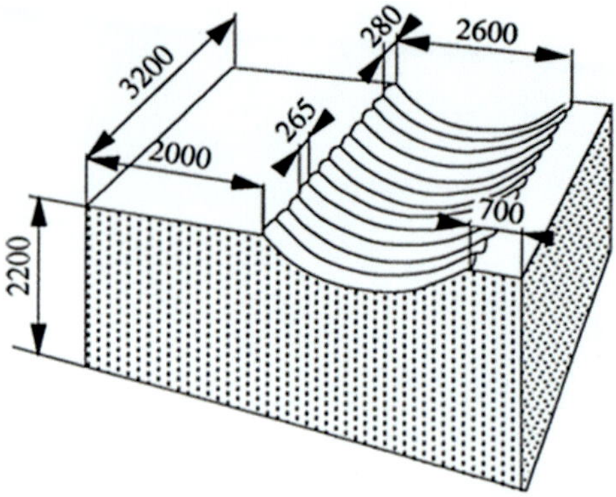

FIGURE 3.26 Illustration of a full-scale milling load case based on full-scale observation. **(Re-produced from Jagodkin, 1963.)**

uniaxial shearing and crushing strength values are used. The model is calibrated with some full-scale shaft measurements together with observations of the milled ice block dimensions. Ignatjev (1964, 1966) developed a model for calculating the required strength of an icebreaker propeller blade.

Noble and Bulat (1981) discuss the existing regulations and results of full-scale shaft measurements on some Canadian Coast Guard icebreakers. They examine the changes of contact geometry of a single blade during passage through an ice block in a milling situation. It is shown that the load may shift from the pressure side to the suction side. This is used to explain the differences between the traditional Jagodkin (1963) model in milling case with the full-scale measurements. The main differences exist on the blade loads, which acts backwards even though the ship is going ahead. Wind (1983) supposed that the interaction can be separated into two parts, i.e., into milling of ice and impact of ice. His proposal for the simplified interaction model is predominantly based on the ice impact.

The model proposed by Kotras et al. (1985) differs from previously introduced ones as it is based on the idea of dividing the propeller into small finite cylindrical strips. This is a practical way to handle the pitch angle, the thickness, the cord length and the champer of the sections that change gradually with the radius. Another important feature of this proposal is that all the running quadrants of the propeller have to be examined, that is, in practice high contact loads over the face or back of the blade.

A more rigorous analytic approach to the failure stress at the face side has been used in the next category of models. Belyashov and Shpakov (1983) and Belyashov (1993) present a method that takes into account the stress state around the blade more accurately. Some tests performed with a propeller blade-like tool and cutting of natural fresh-water ice form the basis for the model. The model is the most theoretically advanced of those published before. The failure mechanism is shear of chip elements to the preceding groove.

Sasajima and Mustamäki (1984) based their method for calculating the milling forces based on model test results. The model is developed for two-dimensional blade segment similar to Belyashov model. Chernuka et al. (1989) have developed a milling model in which the loads for each two-dimensional blade section have been generated for typical contact geometry with FEM calculations being the most extensive of the published models before 1989.

3.4.2 Development towards the Current IACS Ice-Class Rules

Veitch (1992) has assumed in his early model a similar face side failure mechanism as Belyashov and has applied a steady contact pressure along the back side contact length. Veitch (1995) model takes into account the dynamics between the ice block and the propeller and he has used a test series with blade-like tools as well. The pressure distribution on the leading-edge area was analysed computationally with boundary element method. A Hertz-type contact pressure distribution was developed for the leading edge and ice failure is characterized by an ice chipping process. A spherical-shaped ice block is used for the blade-ice dynamic calculations.

Koskinen et al. (1996) and Soininen (1998) have published a method that is in principle similar. The ice block propeller dynamics simulation was performed for a variety of propellers and ice block dimensions and verified with the MS Gudingen full-scale data. Soininen (1998) is using slip-line theory with Mohr-Coulomb failure criterion to study the failure load of solid ice. The pressure distribution due to the extrusion of crushed ice is studied using both viscous and granular models. A simplified method to calculate the effective load is formed for various contact geometries and ice strength parameters. This research formed the basic background of the current propulsion rules as described further in Chapter 6, Section 6.3.

Today, the main research activities are devoted to the numerical modelling of the propeller-ice interaction. As shown in Figure 3.25, this is a complicated research topic. Understanding the ice loads and structural responses during propeller-ice interactions is crucial for optimizing propeller design and ensuring operational safety in ice-prone regions. A summary of the latest numerical modelling of propeller-ice interaction is described in further detail in Chapter 8, Section 8.3.

However, all these numerical models require future validation using good model scale of full-scale data, especially the full-scale data is particularly important, as scaling of the propeller-ice interactions in model scale is a demanding task.

3.5 SUMMARY

Ship-ice interaction includes several different aspects involving ice failure and motion as well as structural deformation. These phenomena couple and compete with each other and demonstrate different levels of importance under different contact scenarios. Factors influencing this process include at least ice mechanical property, ice thickness and floe size, loading speed and ship parameters. On the contact area, crushing and flaking take place, which result in high-pressure zones carrying most of the loads. Away from the contact area, cracks form in the radial and circumferential direction, which either splits the ice floe or breaks it by bending. In dynamic compressive ice fields, ice may fail through buckling at the midship region. These processes break ice into smaller parts or pieces, which continue to interact with the ship at discrete elements. The broken ice can be rotated, submerged, or pushed away by the ship, making space for the ship to advance. From the perspective of structure, the load distributes unevenly on the frames and shell, and spreads to the adjacent frames by the connecting shell. In the later chapters, we will introduce methods to calculate ice loads and resistance, which originate from the mechanism introduced in this chapter.

The propeller-ice interaction is also shortly reviewed, especially from the rule development point of view as described in further detail in Chapter 6. When navigating in ice-covered waters, propellers are subjected to complex ice-water-propeller multiphase coupling effects. Their interaction with sea ice primarily occurs in two forms: indirect and direct interactions. When the sea ice is at a certain distance from the propeller, indirect interactions dominate, with sea ice mainly disturbing the flow field around the propeller. In case of direct interaction, the propeller comes into direct contact with ice blocks, resulting in substantial ice loads on the propeller. The direct interaction between the propeller and ice includes two forms: collision and milling. Of these, milling poses the greatest threat to the propeller. Comparing the hull-ice interaction, the impact speed between propeller and ice is remarkably higher causing different failure mechanisms of the ice as described in this section.

REFERENCES

Belyashov, V. A., 1993. Method for calculating ice loads encountered by propeller blades. In *Proceeding of the 12th International Conference on Port and Ocean Engineering under Arctic Conditions*, Hamburg, 17–20 August 1993.

Belyashov, V. A., and Shpakov, V. S., 1983. On mechanics of ice crushing by propeller blades (trans.). In: *Ice Mechanics and Physics. Nauka, Academy of Sciences of the USSR*, Moscow.

Chernuka, M. W., Jategaongar, R. P., Norwood, M. E., et al., 1989. Development of a procedure for predicting propeller-ice interaction forces. *Transport Canada Publication*, 263–281.

Daley, C. G., 2007. Reanalysis of ice pressure-area relationships. *Marine Technology and SNAME News*, 44(04), 234–244.

Hänninen, S., 2001. *Incidents and accidents in winter navigation in the Baltic Sea, winter 2002–2003*. Winter Navigation Research Board, Report No. 54.

Ignatjev, M. A., 1964. Determination of ice loads encountered by ship propeller blades, (translated from Russian). *Problems of the Arctic and Antarctic*, 15, 41–51.

Ignatjev, M. A., 1966. *Screw propellers of ice going ships*. USSR.

ISO, 2019. Petroleum and natural gas industries—Arctic offshore structures. ISO 19906. https://www.iso.org/standard/65477.html [18.03.2026]

Jagodkin, V. Y., 1963. Analytical determination of the resistance moment of a propeller during its interaction with ice, (translated from Japanese). *Problem of the Arctic and Antarctic*, 13, 79–88.

Jussila, M., and Soininen, H., 1991. *Interaction between ice and propeller*. VTT Technical Research Centre of Finland. Valtion teknillinen tutkimuskeskus. Tiedotteita No. 1281

Kim, J. H., Lu, W., Lubbad, R., Løset, S., and Jang, B. S., 2022. Dynamic bending of an ice wedge resting on a Winkler-type elastic foundation. *Cold Regions Science and Technology*, 199, 103579.

Kõrgesaar, M., Kujala, P., and Romanoff, J., 2018. Load carrying capacity of ice-strengthened frames under idealized ice load and boundary conditions. *Marine Structures*, 58, 18–30.

Koskinen, P., Jussila, M., and Soininen, H., 1996. *Propeller ice load models*. Technical Research Centre of Finland, Research Notes 1739.

Kotras, T., Humphreys, D., Baird, A., Morris, G., and Morley, G., 1985. Determination of propeller-ice milling loads. Proceedings of the Fourth International Offshore Mechanics and Arctic Engineering Symposium, Vol. 2, pp. 336–343, Amer Society of Mechanical, Dallas, July 1985.

Kujala, P., Goldstein, R., Osipenko, N., and Danilenko, V., 1993. *A ship in compressive ice – Analysis of the ice failure process*. Helsinki University of Technology Research Report, M-165*.

Li, F., Kotilainen, M., Goerlandt, F., Kujala, P., 2019. An extended ice failure model to improve the fidelity of icebreaking pattern in numerical simulation of ship performance in level ice. *Ocean Engineering*, 176, 169–183.

Li, F., Kujala, P., and Montewka, J., 2019. A ship in compressive ice: an overview and preliminary analysis. In *Proceedings 25th International Conference Port Ocean Engineering Under Arctic Conditions*, Delft, Netherland.

Lu, W., 2014. *Floe ice–sloping structure interactions*. NTNU.

Masterson, D. M., Frederking, R. M., Wright, B., Kärnä, T., and Maddock, W. P., 2007. A revised ice pressure-area curve. In *Proceedings of the 19th International Conference on Port and Ocean Engineering under Arctic Conditions*, Vol. 1, pp. 305–314, Dalian, China, June 27–30, 2007.

Noble, P., and Bulat, V., 1981. Icebreaker bow forms-A parametric variation. Presented at *Ice Tech '81 sponsored by the Eastern Canadian Section of SNAME*, 16-19 June 1981.

Palmer, A. C., Dempsey, J. P., and Masterson, D. M., 2009. A revised ice pressure-area curve and a fracture mechanics explanation. *Cold Regions Science and Technology*, 56(2–3), 73–76.

Quinton, B. W., 2019. Lateral (sliding) motion of design ice loads on IACS polar classed structures. *Ships and Offshore Structures*, 14(sup1), 281–291.

Sasajima, T., and Mustamäki, E., 1984. Ice-milling load encountered by a controllable pitch propeller. In *Proceedings of the 7th IAHR International Symposium on Ice*, Hamburg, Germany.

Schulson, E. M., and Duval, P., 2009. *Creep and fracture of ice*. Cambridge University Press.

Soininen, H., 1998. *A propeller-ice contact model. dissertation for the degree of doctor of technology*. Helsinki University of Technology, pp. 113–121.

Suominen, M., Kujala, P., Romanoff, J., Remes, H., 2017. Influence of load length on short-term ice load statistics in full-scale. *Marine Structures*, 52, 153–172.

Taylor, R. S., Richard, M., and Hossain, R., 2019. A probabilistic high-pressure zone model for local and global loads during ice-structure interactions. *Journal of Offshore Mechanics and Arctic Engineering*, 141(5), 051604.

Timco, G. W., and Sudom, D., 2013. Revisiting the Sanderson pressure–area curve: Defining parameters that influence ice pressure. *Cold regions Science and Technology*, 95, 53–66.

Valtonen, V., Bond, J., and Hindley, R., 2020. Improved method for non-linear FE analysis of Polar Class ship primary structures. *Marine Structures*, 74, 2020.

Varsta, P., 1983. *On the mechanics of ice load on ships in level ice in Baltic Sea*. Doctoral dissertation, VTT Technical Research Centre of Finland.

Varsta, P., and Riska, K., 1977. Failure process of ice edge caused by impact with ship's side. In *From the Symposium on Ice, Ships and Winter Navigation*, Oulu University, 16–17 December 1977.

Veitch, B. J., 1992. *Propeller ice interaction*. PhD Thesis. Faculty of Mechanical Engineering, Helsinki University of Technology, Finland.

Veitch B., 1995. Predictions of ice contact forces on a marine screw propeller during the propeller-ice cutting process. *Acta Polytechnica Scandinavica Mechnical Engineering*. Helsinki, 140 p.

Wind, J., 1983. The dimensioning of high power propeller system for arctic ice-breaking vessels. In *Proceedings of the 5th Lips Propeller Symposium*. Drunen, The Netherlands, May 1983.

Xu, Y., Kujala, P., Hu, Z., Li, F., and Chen, G., 2020. Numerical simulation of level ice impact on landing craft bow considering the transverse isotropy of Baltic Sea ice based on XFEM. *Marine Structures*, 71, 102735.

Zhou, L., 2012. *Numerical and experimental investigation of stationkeeping in level ice*. Elsevier B.V.

4 Ice Loading on Ship Hull

4.1 ENERGY-BASED METHOD

Popov et al. (1967) proposed an ice-load calculation method based on energy conservation, which has a profound effect to the current polar-class regulations after further development by Daley (1999). In this section, we will introduce this method in detail. The local ship-ice interaction will be the focus area as this forms the basis of the current ice-class rules. On heavy ice, the ramming of ship against the ice can also cause remarkable global loads causing bending of the whole ship beams; this is not described here, but if readers are interested in this topic, we can recommend to get familiar with the work of Riska (1987).

4.1.1 Assumptions

Popov assumes two ship-ice contact scenarios, schematically shown in Figure 4.1. In the first scenario, a ship collides with an ice floe with finite size, which gains kinetic energy after the collision. In the second scenario, a ship collides with a large ice sheet which does not move after the collision. The basic assumptions of Popov method include (see also Figure 4.2):

- The ship is symmetric about its midplane, and the centre of gravity is located on the midplane;
- The ice floe is in circular shape, with thickness much smaller than its radius;
- The hydrodynamic force is much smaller than ice contact force;
- The translation of the ship and ice is small during the interaction;
- The friction force is negligible in the interaction process;
- The ship body is rigid, therefore does not deflect; the ice-crushing process is completely plastic (no elasticity considered);
- The water surface is calm, i.e., there is no wave.

Before the contact, the ship moves in *x*-direction with a constant speed of u_0, while the ice floe is set to be static. Popov did not account for the energy gained from propulsion system. For the first scenario where the ship collides with ice floe, Eq. (3.1) can be written now as

$$\Delta E^S_{\text{kinetic}} = E^I_{\text{failure}} + E^I_{\text{kinetic,ice}} \tag{4.1}$$

where $\Delta E^S_{\text{kinetic}}$ is the loss of kinetic energy by the ship. So, the loss of ship's kinetic energy is partly used to break ice by crushing, and partly converted to the kinetic energy of the ice floe.

DOI: 10.1201/9781003676829-4

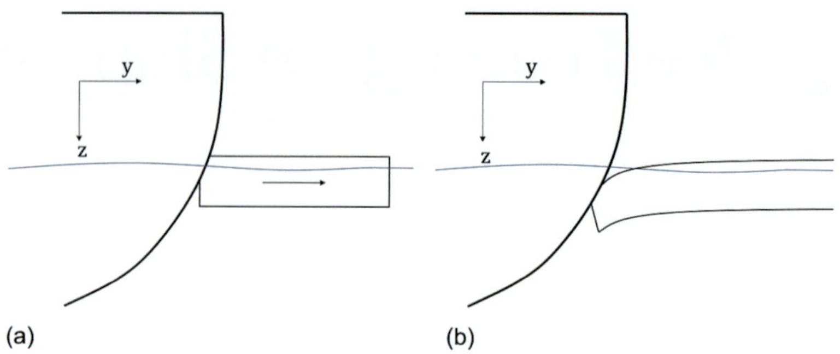

FIGURE 4.1 Contact scenario, (a) ice floe; (b) ice sheet.

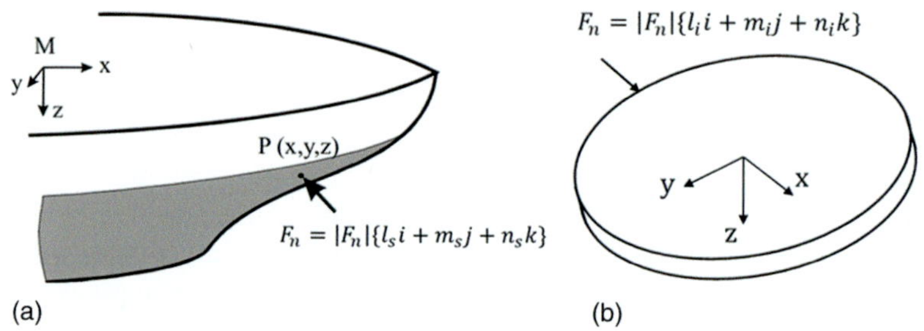

FIGURE 4.2 The coordinate system and the location of contact.

For the second scenario where the ship collides with an ice sheet, Eq. (3.1) can be written now as

$$\Delta E^S_{\text{kinetic}} = E^I_{\text{failure}} + E^I_{\text{potential}} \tag{4.2}$$

Therefore the loss of ship's kinetic energy is partly used to break ice by crushing, and partly converted to the potential energy of the ice floe as the result of deflection.

4.1.2 Reduced Mass and Velocity

Ship-ice contact is a six-degree-of-freedom (6-DoF) problem, which is difficult to solve analytically. To simplify the calculation process, Popov proposed the concept of reduced mass so the problem can be reduced to 1 DoF.

Denote S as the change of momentum in the direction normal to the contact area. The change of momentum in surge, sway, heave, roll, pitch and yaw motion is then the magnitude of S project to each DoF, i.e., the right-hand side of Eq. (4.3). This equals the product of the mass or rotational inertia (including added mass) and the corresponding velocity or angular velocity change, i.e., the left-hand side of Eq. (4.3). Therefore, for the ship with 6 DoF motion, the following equations hold:

$$
\begin{aligned}
M_s(1+a_{1s})(u_s-u_0) &= -l_sS \\
M_s(1+a_{2s})v_s &= -m_sS \\
M_s(1+a_{3s})w_s &= -n_sS \\
I_{xs}(1+a_{4s})\phi_s &= -\lambda_sS \\
I_{ys}(1+a_{5s})\theta_s &= -\mu_sS \\
I_{zs}(1+a_{6s})\psi_s &= -\eta_sS
\end{aligned}
\tag{4.3}
$$

where M_s, I_{xs}, I_{ys}, I_{zs} are the mass and moments of inertia of the ship in different directions; a_{is} is the added mass coefficient; u_s, v_s, w_s, ϕ_s, θ_s, ψ_s are the speed components after the collision; l_s, m_s, n_s are the projections of the unit vector perpendicular to the contact surface onto x-, y- and z-axes; λ_s, μ_s, η_s are the arms of the momentum in x-, y- and z-directions, as per

$$
\begin{aligned}
\lambda_s &= yn_s - zm_s \\
\mu_s &= zl_s - xn_s \\
\eta_s &= xm_s - yl_s
\end{aligned}
\tag{4.4}
$$

where (x, y, z) is the location of collision in the ship coordinate system.

Based on the theorem of momentum conservation, S is also the gain in momentum of the ice floe. For the ice floe in 3 DoF motion, the following equation holds:

$$
\begin{aligned}
M_i(1+a_{1i})u_i &= l_iS \\
M_i(1+a_{3i})w_i &= n_iS \\
I_{yi}(1+a_{5i})\theta_i &= n_iRS
\end{aligned}
\tag{4.5}
$$

where the meanings of the symbols are similar to those for the ship motion, hence not repeated here.

Based on the sixth assumption in Section 4.1.1, after the collision, the ship and ice have the same speed in the direction perpendicular to the contact surface. According to the theory of rigid body motion, the following equation holds:

$$
\begin{aligned}
&l_su_s + m_sv_s + n_sw_s + (\theta_sz - \psi_sy)l_s + (\psi_sx - \phi_sz)m_s + (\phi_sy - \theta_sx)n_s \\
&= l_iu_i + n_iw_i - R\theta_in_i
\end{aligned}
\tag{4.6}
$$

Note that $n_s = n_i$, the above three groups of equation can be resolved to yield

$$
S = \frac{\dfrac{M_s}{C_s}\dfrac{M_i}{C_i}u_0l_s}{\dfrac{M_s}{C_s} + \dfrac{M_i}{C_i}}
\tag{4.7}
$$

where

$$C_s = \left(\frac{l_s^2}{1+a_{1s}} + \frac{m_s^2}{1+a_{2s}} + \frac{n_s^2}{1+a_{3s}}\right) + \left(\frac{\lambda_s^2}{(1+a_{4s})I_{xs}} + \frac{\mu_s^2}{(1+a_{5s})I_{ys}} + \frac{\eta_s^2}{(1+a_{6s})I_{zs}}\right)M_s$$

$$C_i = \left(\frac{l_i^2}{1+a_{1i}} + \frac{n_i^2}{1+a_{3i}}\right) + \frac{n_i^2 R^2}{(1+a_{5i})I_{yi}} M_i$$

Let $M_{s,red} = \frac{M_s}{C_s}$, $M_{i,red} = \frac{M_i}{C_i}$, $u_{red} = u_0 l_s$, Eq. (4.7) can be rewritten as

$$S = \frac{M_{s,red} M_{i,red}}{M_{s,red} + M_{i,red}} u_{red} \tag{4.8}$$

The problem is the equivalent to two objects with mass $M_{s,red}$ and $M_{i,red}$ colliding with each other with the restitution coefficient of 0 (centre of mass in line with the speed of collision), which is simplified into a 1-DoF problem. $M_{s,red}$ and $M_{i,red}$ are therefore termed as the reduced mass of the ship and ice.

When the mass of the ice floe is infinite, Eq. (4.8) becomes

$$S = M_{s,red}\, u_{red} \tag{4.9}$$

which is then the momentum change when a ship collides with an ice sheet with infinite size. From this equation, it can be noticed that the momentum of the ship in the direction perpendicular to the contact surface becomes zero after the collision.

The above derivations assume no effect from waves. With the presence of ocean waves, the hydrodynamic coefficients including added mass and damping become related to the magnitude and frequency of the wave. The velocity at collision will differ from the reduced velocity calculated as above. These facts can result in differences in the eventually calculated ice loads. A recent doctoral thesis by Jiang (2024) investigated this issue using potential flow theory. Readers who are interested in this topic are referred to this thesis for further information.

4.1.3 Ship Colliding Ice Floe

In the previous subsection, the problem of ship-ice collision is simplified into a 1-DoF collision problem in the direction perpendicular to the contact area. As illustrated in Figure 4.3, ***N*** is the direction perpendicular to the contact surface; t_0 is the time instant at the initiation of the collision, t_1 is an instant during the collision; x_1 and x_2 are the translations of the ship and the ice floe at t_1 along ***N***-direction; ζ is the crushing indentation. Based on the geometrical relationship, there holds

$$\zeta = x_1 - x_2 \tag{4.10}$$

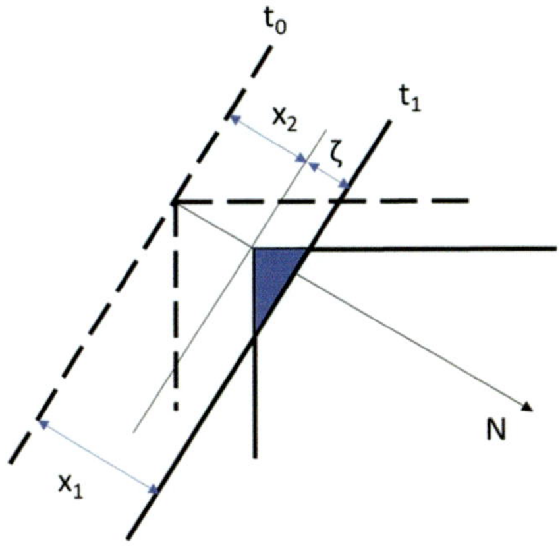

FIGURE 4.3 Translation of ship and ice during the interaction process.

$P(\zeta)$ denotes the contact force at crushing indentation ζ. Based on Newton's second law, there holds

$$P(\zeta) = -M_{s,red}\ddot{x_1} = M_{i,red}\left(\ddot{x_1} - \ddot{\zeta}\right) \tag{4.11}$$

Eliminating $\ddot{x_1}$ with Eq. (4.8) yields

$$\frac{M_{i,red}M_{s,red}}{M_{i,red} + M_{s,red}}\ddot{\zeta} = -P(\zeta) \tag{4.12}$$

Now consider the relationship between the contact force P and the crushing indentation ζ. The magnitude of P can be calculated as the product of the contact area and the nominal pressure on this area. Typically, A could be expressed as a function of ζ by

$$A = C\zeta^a \tag{4.13}$$

where C and a are constants depending on the edge geometry of ice, see Daley et al. (1998). Popov assumes that the nominal pressure p equals the crushing strength of ice, denoted by σ_c. Let $M_{red} = \frac{M_{i,red}M_{s,red}}{M_{i,red} + M_{s,red}}$, with Eq. (4.13), Eq. (4.12) can be written as

$$M_{red}\ddot{\zeta} = -C\sigma_c\zeta^a \tag{4.14}$$

which is a differential equation regarding ζ, with its initial value $\zeta(0)$ as

$$\zeta(0) = 0, \dot{\zeta}(0) = u_{red}$$

Here we do not show the derivation process of Eq. (4.14). Noticing that the relative speed between ship and ice after collision is 0, i.e. $\dot{\zeta} = 0$, we get

$$\zeta_{max} = \left[\frac{(a+1)M_{red}u_{red}^2}{2C\sigma_c}\right]^{\frac{1}{1+a}} \tag{4.15}$$

which gives the maximum ice load P_{max} as

$$P_{max} = C\sigma_c\left[\frac{(a+1)M_{red}u_{red}^2}{2C\sigma_c}\right]^{\frac{a}{1+a}} \tag{4.16}$$

The full time-history of the ice load as well as its maximum value can then be obtained.

4.1.4 Ship Colliding Thick Ice Sheet

As illustrated in Figure 4.4, when ship collides with thick ice, the ice sheet does not move, but instead deforms due to the downward bending force. When the ship speed along the direction normal to the contact surface drops to 0, the consumed ship kinetic energy is all converted into the strain energy of ice bending and the failure energy through crushing, as mentioned earlier in Eq. (4.2). The terms in this equation can be written as:

$$\Delta E_{\text{kinetic}} = \frac{1}{2}M_{s,red}u_{red}^2$$

$$E_{\text{failure}} = \int_0^{\zeta_{max}} P d\zeta = C\sigma_c \frac{\zeta_{max}^{a+1}}{a+1}$$

$$E_{\text{potential}} = \frac{1}{2} d P_{max} \sin\beta \tag{4.17}$$

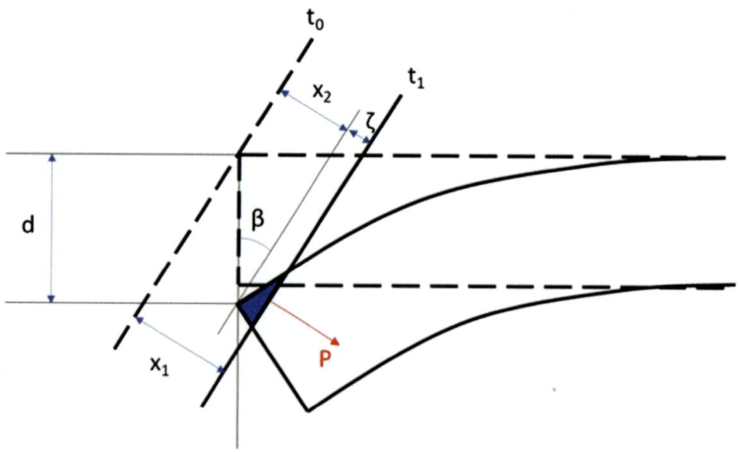

FIGURE 4.4 Translation of ship and ice during the interaction process.

where d is the deflection of the ice edge under the vertical component of the contact force, which, according to the bending theory of floating semi-infinite plate, can be written as:

$$d = \frac{P_{max} \sin \beta}{2\sqrt{\rho g D}} \tag{4.18}$$

where ρ is the density of water, g the gravitational acceleration and D the bending rigidity. Combining Eqs. (4.2, 4.17, 4.18) yields

$$\frac{1}{2} M_{s,red} u_{red}^2 = C\sigma_c \frac{\zeta_{max}^{a+1}}{a+1} + \frac{P_{max}^2 \sin^2 \beta}{4\sqrt{\rho g D}} \tag{4.19}$$

As $P_{max} = C\sigma_c \zeta_{max}^a$, we can get ζ_{max} as per

$$\zeta_{max} = \left(\frac{M_{s,red} u_{red}^2}{\frac{2\sigma_c C}{a+1} + \frac{\sigma_c^2 C^2 \sin^2 \beta}{2\sqrt{\rho g D}} \zeta_{max}^{a-1}} \right)^{\frac{1}{a+1}} \tag{4.20}$$

which gives the maximum ice load P_{max} as

$$P_{max} = C\sigma_c \left(\frac{M_{s,red} u_{red}^2}{\frac{2\sigma_c C}{a+1} + \frac{\sigma_c^2 C^2 \sin^2 \beta}{2\sqrt{\rho g D}} \zeta_{max}^{a-1}} \right)^{\frac{a}{a+1}} \tag{4.21}$$

In this equation, ζ_{max} appears on both sides of the equation, therefore should be solved iteratively. One can use $\zeta_{max} = 0$ as the initial value and iterate several steps to approximate ζ_{max}.

4.1.5 Flexural Limit

The above derivation only accounts for crushing failure of ice without considering bending failure. In reality, if the ice floe/sheet is not very thick, under the vertical component of the contact force, ice may fail by bending before the relative speed drops to 0. Popov uses an empirical formula to calculate the vertical force resulting in ice flexural failure, expressed as:

$$P_{flex} = 0.52\sigma_b h^2 \tag{4.22}$$

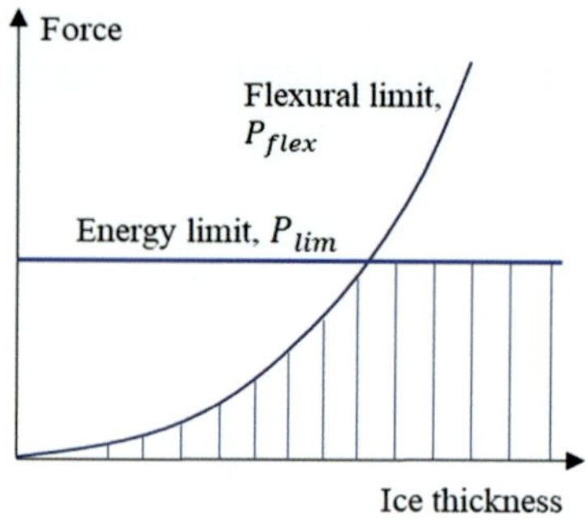

FIGURE 4.5 The competition between energy limit force and flexural limit force.

where σ_b is the flexural strength, h the ice thickness. According to Figure 4.4, the magnitude of the vertical force can be calculated as $P \sin \beta$, thereby resulting in

$$P_{lim} = \frac{0.52\sigma_b h^2}{\sin \beta} \tag{4.23}$$

The maximum ice load is then the smaller value between P_{lim} and P_{max}. This can be illustratively sketched in Figure 4.5.

4.1.6 Application of Popov Method in Polar Rules

The Popov method laid the foundation for the International Association of Classification Societies (IACS, 2016) ice-class ship structural design regulations. These regulations adopt a general wedge-shaped collision between the ship and the ice layer as the design scenario. The Popov method has been simplified in these regulations, disregarding the strain potential energy absorbed by ice deformation. Specifically, the regulations assume $E_{\text{potential}} = 0$ in the equation, implying that the kinetic energy loss is entirely used for the compression and crushing of ice.

$$\Delta E_{\text{kinetic}} = E_{\text{failure}} \tag{4.24}$$

Based on this, calculations for the magnitude of ice loads and contact area are performed. Further details about the IACS ice-class ship structural regulations will be provided in Chapter 6.

4.2 WEDGE BENDING SOLUTION

Popov's energy-based method primarily deals with ship's interaction with thick ice, where crushing dominates the whole process. The flexural failure, however, is extensively simplified as indicated by Eq. (4.22). When ships navigate through relatively thin ice in continuous mode, ice breaks by flexure and limit the ice loads as illustrated in Figure 4.5. A more rigorous derivation of the ice loads under such condition is needed. Nevel (1961) solved the ice bending problem under quasi-static loading by theoretical derivation assuming the ice as a narrow wedge. Although this method has not been incorporated into ice-class rules, it has received lots of attention in the academia and been used to develop ship performance models. In addition, the derivation process contains important aspects which forms the basis for many semi-empirical formulae. Therefore, in this section, we make an introduction of Nevel's method.

4.2.1 Assumptions

Nevel introduces the concept of narrow wedge which is shaped as shown in Figure 4.6. By 'narrow' it means that the wedge opening angle is so small that the stress only varies along the radial direction and remains constant along the circumferential direction. As a wedge, the width along the radial direction, denoted as b, is a linear function of its radial coordinate, denoted as x, i.e., $b(x) = b_0 x$, where k is a coefficient. The basic assumptions of Nevel's narrow wedge solution includes:

1. Ice is a perfectly elastic material with no viscosity, i.e., the loading rate has no effect on the material property. This is justified by the relatively high strain rate during ice-structure interaction;
2. Ice is a brittle material, which cracks immediately when the tensile stress reaches the tensile strength;
3. The water can be treated as an elastic foundation;
4. The neutral axis is located at half the height of the beam;
5. The ice wedge angle is small and there is no variation of stress and strain in the angular direction;
6. Assumptions of Euler-Bernouli beam theory hold and there is no in-plane force; and
7. The loading process is quasi-static.

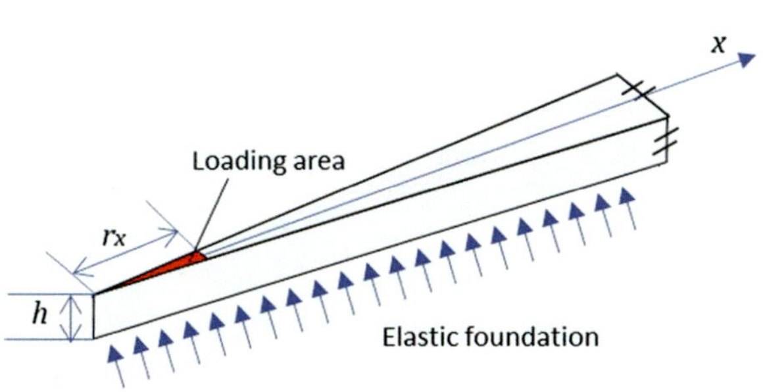

FIGURE 4.6 Nevel's narrow wedge theory.

4.2.2 Derivation

Following the basic assumptions of Euler-Bernoulli beam and assuming the water as an elastic foundation, the moment acting on a cross-section of the narrow wedge at position x can be expressed as:

$$M = EI\frac{d^2y}{dx^2} = E\frac{b(x)h^3}{12}\frac{d^2y}{dx^2} \tag{4.25}$$

where y is the deflection at position x, E is the elastic modulus of ice, h is the thickness of ice and I the moment of inertia. With $b(x) = b_0x$, Eq. (4.25) can be expressed as:

$$M = E\frac{b_0xh^3}{12}\frac{\mathrm{d}^2y}{\mathrm{d}x^2} \tag{4.26}$$

The equilibrium on an infinite small section of the narrow wedge is then established

$$\frac{\mathrm{d}^2}{\mathrm{d}x^2}\left(\frac{Ekxh^3}{12}\frac{\mathrm{d}^2y}{\mathrm{d}x^2}\right) + \rho gyb_0x = q(x) \tag{4.27}$$

where $q(x)$ is the load acting on the wedge at position x in N/m, which is 0 away from the loading area, and proportional to x on the loading area, i.e. $q(x) = q_0x$; ρ is the density of water and g the gravity acceleration. Here the first term is the internal moment within the wedge, and the second term is the hydrostatic force due to the deflection of the ice wedge in the vertical direction.

Further derivation from Eq. (4.27) gives

$$\frac{Eh^3}{12\rho g}\frac{d^4y}{dx^4} + 2*\frac{Eh^3}{12\rho g}\frac{1}{x}\frac{d^3y}{dx^3} + y = \frac{q_0}{b_0\rho g} \tag{4.28}$$

Let us now introduce an important parameter, named by the characteristic length, denoted by l_c as per

$$l_c = \sqrt[4]{\frac{Eh^3}{12\rho g}} \tag{4.29}$$

And let $\chi = \frac{x}{l_c}$, which is a non-dimensional coordinate. Eq. (4.28) can be now rewritten as:

$$\frac{d^4y}{d\chi^4} + \frac{2}{\chi}\frac{d^3y}{d\chi^3} + y = \frac{q_0}{b_0\rho g} \tag{4.30}$$

Now the equation is non-dimensional in terms of the coordinate and free of thickness and elastic modulus parameters. The solution can be obtained by solving this differential equation. Nevel derived closed-form solutions to this equation. The form of this solution is lengthy and hence not provided here. Readers interested in this topic can refer to Nevel's paper. The eventual solution is in the form of

$$y(\chi) = A*du_1(\chi) + B*du_2(\chi) + C*du_3(\chi) + D*du_4(\chi) \tag{4.31}$$

where $du_1(\chi)$, $du_2(\chi)$, $du_3(\chi)$ and $du_4(\chi)$ are specific functions of χ; A, B, C, D are the unknowns to be solved from boundary conditions.

It should be emphasized here that l_c is the key parameter generalizing the solution to any thickness and elastic modulus. The characteristic length l_c is therefore a very important factor widely used in ice engineering problems.

Once the solution to this equation is obtained, the tensile stress on the upper edge of the beam incurred by the bending moment can be calculated by

$$\sigma_{\text{wedge}} = \frac{M}{W} = \frac{Eh}{2}\frac{d^2y}{dx^2} \tag{4.32}$$

where $W = \frac{b(x)h^2}{6}$ is the section modulus. Bending failure is initiated when σ_{wedge} reaches the flexural strength σ_f, i.e.

$$\sigma_{\text{wedge}} = \sigma_f \tag{4.33}$$

Following the above derivation, the limiting load resulting from flexural failure can be obtained.

4.2.3 Demonstration

Next, a demonstration of Nevel's wedge solution is given. Assume an ice wedge with wedge angle 60°, thickness 0.6 m and elastic modulus 2 GPa, loaded at the tip vertically with pressure 1 MPa. The deflection and stress field along the wedge beam with different load depth r (i.e. r_x in Figure 4.7) are calculated and plotted in Figure 4.7. The curves of tensile stress feature an increase first followed by the decrease, resulting in a maximum tensile stress at some distance away from the contact region. The location of the maximum stress is then the location where the crack most likely occurs.

The location and magnitude of the maximum tensile stress are extracted from Figure 4.7 and then plotted in Figure 4.8 as functions of the load depth r. When a ship contacts an ice wedge, the edge of the ice gets crushed and flaked gradually, with the load depth r increasing. The maximum tensile stress increases accordingly until it reaches the tensile strength, which is often approximated by the flexural strength. A crack then occurs at the location of the maximum tensile strength, which can be found with the curves in Figure 4.8. Assuming the tensile/flexural strength to be 500 kPa, it is found that the ice will break when the load depth reaches 0.44 m, which corresponds to vertical force of 83.8 kN on the contact area. When the ice breaks, an ice piece of 4.1 m will be broken off the intact ice sheet.

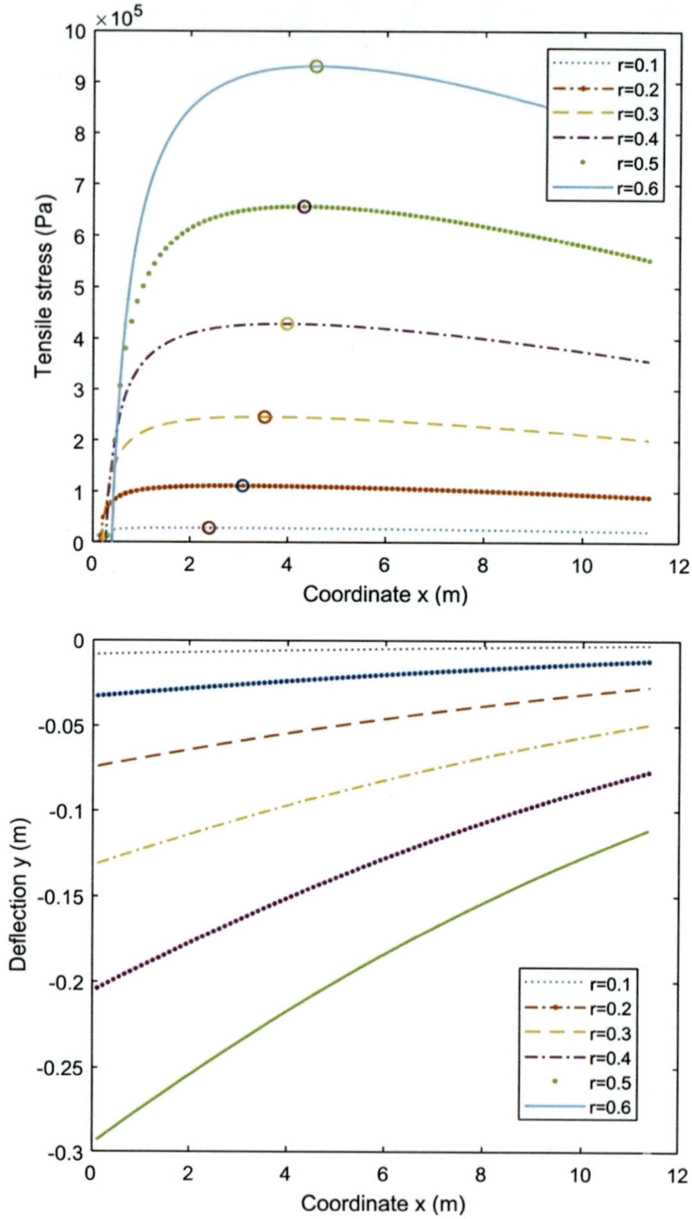

FIGURE 4.7 Stress (upper) and deflection (lower) field along the wedge beam with different load depth; the circles denote the locations of maximum tensile stress.

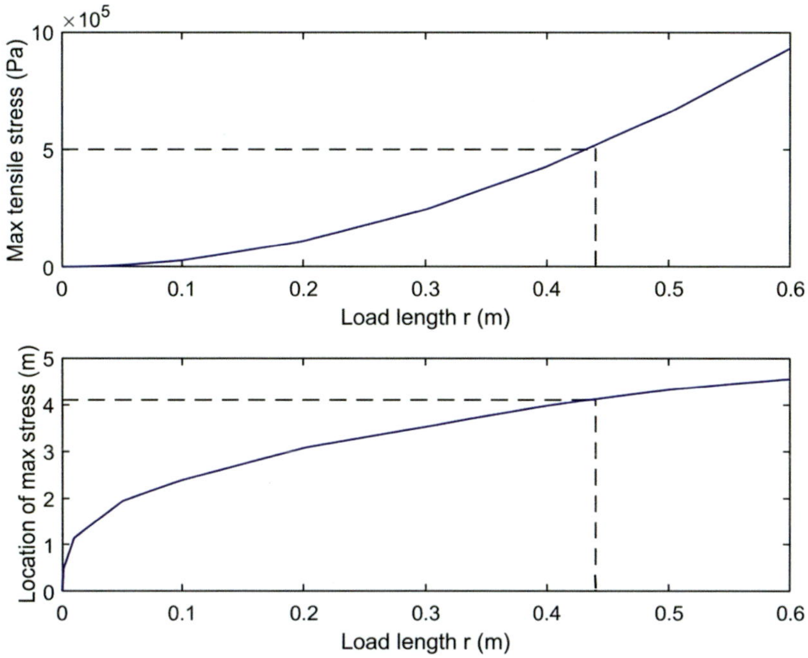

FIGURE 4.8 The magnitude and location of the maximum tensile stress as functions of the load depth r.

4.3 PROBABILISTIC METHODS

Theoretical methods are primarily based on the basic principles of mechanics. Such methods are advantageous in the capability to consider multiple affecting factors via first principle-based derivations. However, the mechanical behaviour of ice is extremely complex as introduced in Chapter 2. It is therefore unavoidable to make necessary assumptions and simplifications to enable theoretical derivation. This leads to large deviations and uncertainties in the resulting ice-load calculation.

As an alternative, one may consider to utilize the information contained in ice-load measurement obtained onboard ships in real ice conditions. The methods to measure ice loads will be introduced later in Chapter 7. Such measurements represent the most credible source of ice loads on ships. However, as will be shown soon, ice loads obtained from full-scale measurements are highly random. In addition, measurements of the prevailing ice condition contain significant uncertainty. Probabilistic methods are hence needed to establish ice-load estimation models based on extensive measurement data.

4.3.1 Principles

4.3.1.1 Source of Randomness

Ice loading on ships is a highly random process. Figure 4.9 presents a 30-minute section of measured ice loads on the hull of a ship sailing in an ice field. One can clearly see the randomness of the load peaks, varying from several kilo Newtons to several hundred-kilo Newtons within a relatively short distance. The randomness comes from various sources, including the geometrical and mechanical features of sea ice, the process of ice failure, and the variation in the contact location. The randomness in the ice loads is then the joint result of the above factors, elaborated below:

- *Randomness in geometrical features*. Ice is formed through natural freezing and has often endured significant deformation during the growing process, resulting in strong stochasticity in thickness, shape, floe size and topology.
- *Randomness in mechanical properties*. The strain modulus, compressive strength, tensile strength and fracture toughness of ice are highly variable locationally and play an important role in the limiting mechanism of peak force.
- *Randomness in ice failure*. Due to the brittleness of ice as a material, the contact with structures is exerted through randomly formed high-pressure zones (HPZs). The number of HPZs and magnitudes of high pressures are stochastic.
- *Randomness in contact location*. When a ship collides with an ice floe, the location of the contact can be random and may or may not hit a local structure member under consideration. Such kinds of stochasticity can also be incurred as the result of crew manoeuvring.

4.3.1.2 Process of Probabilistic Modelling

When we have gathered ice-load measurement data exemplified in Figure 4.9, where high randomness exists, typically, we wish to characterize the random loads with established probability distributions, and find the link between the load distribution and the prevailing ice conditions and ship parameters. In this way, a probabilistic

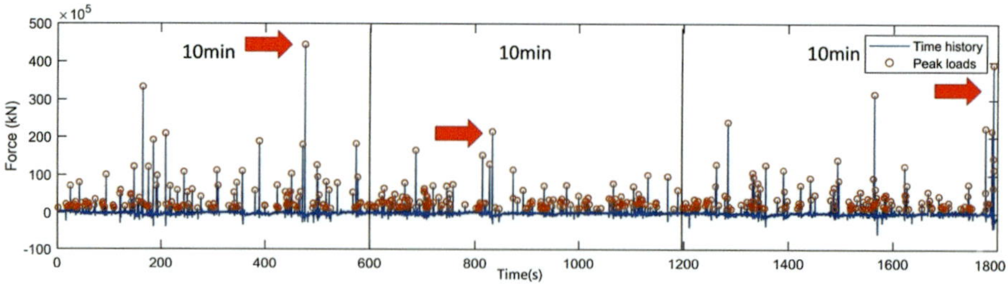

FIGURE 4.9 Time history of measured ice loads and identified load peaks, along with the 10-minute maxima (Li et al., 2021).

model can be established which we can use to make estimations of ice loads on ship hull. The process of probabilistic modelling is shown in Figure 4.10.

Statistical inference needs to be conducted to identify the best probability distribution which can fit the measured ice loads. Researchers have made extensive attempts to find the proper probability distribution to characterize random ice loads. Figure 4.11 gives an example of Weibull, exponential and lognormal distribution fitted to the measured ice loads onboard a polar research vessel. The distribution fitting is conducted via the Maximum Likelihood Estimation (MLE) method. It can be visually judged that Weibull distribution can fit the dataset quite well. To quantify the goodness of fit, one can adopt techniques such as the q-q plot, p-p plot, k^2 test, Kolmogorov–Smirnov test, etc.

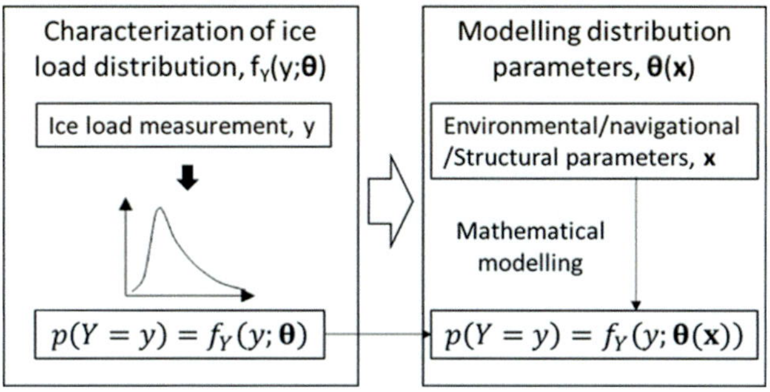

FIGURE 4.10 Process of probabilistic modelling.

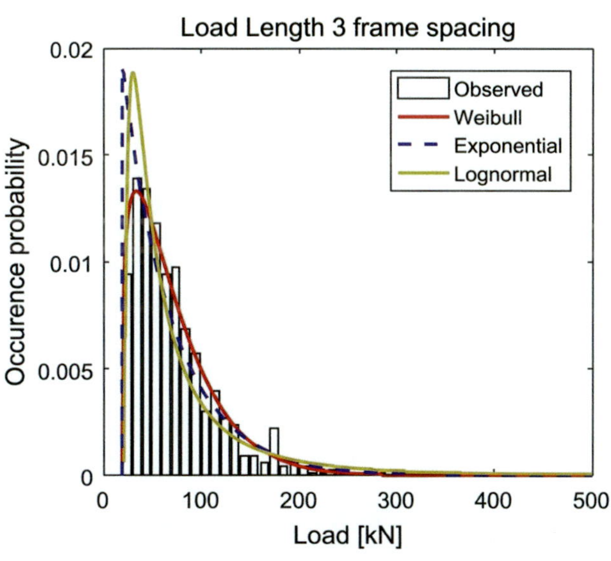

FIGURE 4.11 Example of distribution fitting onto the ice-load dataset (Suominen et al., 2017).

The existing research generally shows that Weibull distribution outperforms other standard distributions on the fitting of individual force peaks exemplified in Figure 4.11. Some non-standard distributions, including truncated distribution and mixture distributions, can fit the force peaks better than Weibull distribution, especially at the tail region where large loads locate. However, the above conclusions do not necessarily apply to other types of ice load. For example, for load maxima of fixed time intervals, the 10-minute maxima as exemplified in Figure 4.9, extreme-value distributions such as the Gumbel distribution often perform well in charactering the random loads.

If we denote Y as the measured ice loads, with statistical inference, we find the best probability distribution, denoted by f_Y, to describe the random ice loads, as per

$$p(Y = y) = f_Y(y; \boldsymbol{\theta}) \tag{4.34}$$

where $\boldsymbol{\theta}$ denotes the distribution parameters and $f_Y(*)$ denotes the *probability density function* (PDF) of ice loads; $p(Y = y)$ denotes the probability density at y.

The essential part of probabilistic methods is to mathematically model the probability distribution parameters, $\boldsymbol{\theta}$, as functions of the prevailing parameters, $\mathbf{x}$, in the dataset, so that the ice-load model can be expressed as:

$$p(Y = y) = f_Y(y; \boldsymbol{\theta}(\mathbf{x})) \tag{4.35}$$

Here $\mathbf{x}$ represents possible factors affecting ice loads, such as ice thickness, ship hull angle, etc. We can see that the core of the probabilistic methods is the mathematical modelling of $\boldsymbol{\theta}(\mathbf{x})$ with measurement data. The quality of the model is then largely dependent on the quality of the data.

4.3.1.3 Extreme Ice-Load Prediction

Now the probabilistic model of ice loads is established. The next question is then how to use this model. Typically, for the design of ship structures, one wants to know what is the maximum load the ship will encounter during its lifetime, e.g., 25 years. This is termed as the *extreme ice load*. Since the individual loads are random variables, the extreme ice load is also a random variable.

For the prediction of extreme loads, a *parent distribution* (or *initial distribution*) is needed to model ice loads of shorter terms. As shown in Figure 4.12, the parent distribution is denoted by f_Y, which models the individual loads in shorter terms,

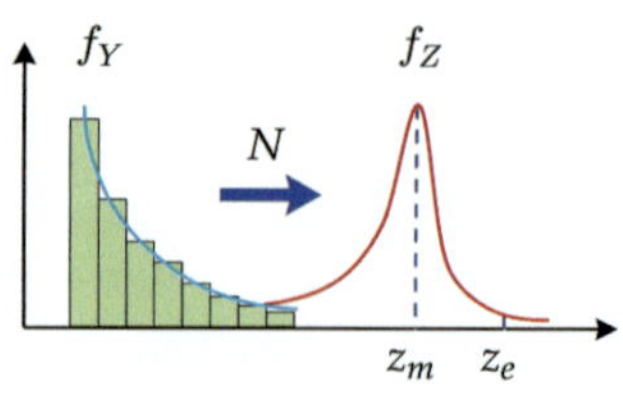

FIGURE 4.12 Illustration of the derivation from the PDF of parent distribution to extreme distribution (Li et al., 2021).

denoted by Y. The extreme ice load during a longer term, e.g., the lifetime of a structure, denoted by Z, is then the maximum of the individual loads, i.e.,

$$Z = max\,(Y_1, Y_2, \ldots Y_N) \tag{4.36}$$

where N is the total number of ice loads which happens during the lifetime of a ship. The cumulative distribution function (CDF) of Z is then derived by

$$\begin{aligned} F_Z(z) &= P(Z < z) = P(Y_1 < z, Y_2 < z, \ldots, Y_N < z) \\ &= P(Y_1 < z)P(Y_2 < z)\ldots P(Y_N < z) = F_Y(z)^N \end{aligned} \tag{4.37}$$

where F_Y is the CDF of individual ice loads, i.e., the parent distribution; $F_Z(z)$ is the CDF of the *extreme value distribution*. By taking the derivative over z, we can get the probability density function of z as:

$$f_z(z) = \frac{d[F_Z(z)]}{dz} = NF_Y(z)^{N-1}\frac{d[F_Y(z)]}{dz} = NF_Y(z)^{N-1}f_Y(z) \tag{4.38}$$

The resulting f_z is schematically shown in Figure 4.12. According to the extreme value theory, when N becomes sufficiently large, $f_z(z)$ will converge to either of the three types of extreme value distributions, namely the Gumbel distribution, the Fréchet Distribution or the Weibull distribution.

Typically, the designers are interested in the most likely extreme load, z_m, and the extreme load corresponding to a probability of exceedance, e.g., 0.01, denoted by z_e. As can be observed, the probability to encounter an extreme load larger than z_m is still considerable. The use of z_e is then a safer choice which has low probability to be exceeded.

Another way of analysing the extreme load is through the *return period*. The return period is defined as *the estimated average time between extreme events*. Here the *extreme events* refer to ice-loading events exceeding a given load level, denoted by z. As can be imagined, smaller loads are expected to happen more often than larger loads during the same time period. This means that larger load levels correspond to longer return periods.

How to calculate the return period given a load level? If we denote n as the average number of ice loads per unit time, usually a year. Following Eq. (4.37), the CDF of the annual maximum ice load is then

$$F_Z(z) = F_Y(z)^n \tag{4.39}$$

The probability of exceedance, denoted by p_e, is defined by

$$p_e = 1 - F_Z(z) = 1 - F_Y(z)^n \tag{4.40}$$

It can be observed that p_e represents the probability that the load level z will be exceeded during one year. The expected time between extreme events, i.e., the return period in years, denoted by T, is then

$$T = \frac{1}{p_e} = \frac{1}{1 - F_Y(z)^n} \text{ years} \tag{4.41}$$

We can then plot z as a function of T to find the load level corresponding to a return period. An example of such a plot is given in Figure 4.15. In the design of large complex structures, the structures are often required to resist loads which occur every 100 years or longer.

4.3.2 Event-Maximum Method

4.3.2.1 Jordaan et al. (1993) Method

The *Event-Maximum Method (EMM)* was proposed by Jordaan et al. (1993) for the estimation of extreme ice loads on ships and offshore structures. The essential part of this methodology is the definition of *event*. Jordaan et al. defined an event as one ship ramming into thick ice, i.e., the process from the ship entering ice until being stopped by ice. The *event-maxima* are then defined as the maximum load during each ramming process. The subsequent modelling is then based on the extracted event-maxima.

Following the notation in Section 4.2.1, the event-maxima are denoted as $Y = (Y_1, Y_2, \ldots Y_N)$. To characterize the randomness in the event-maxima, Jordaan et al. proposed to use exponential distribution to fit the upper tail, as exemplified in Figure 4.13. The probability density function (PDF) of exponential distribution can be expressed by

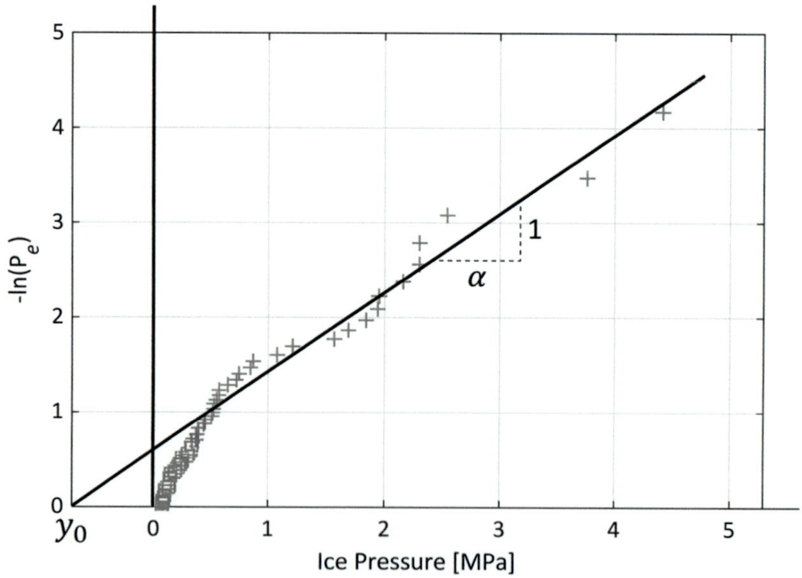

FIGURE 4.13 Linear regression to the load data obtained on the same area (Shamaei et al., 2020).

$$f_Y(y) = \frac{1}{\alpha} \exp\left(-\frac{y - y_0}{\alpha}\right) \tag{4.42}$$

The cumulative distribution function (CDF) is

$$F_Y(y) = 1 - \exp\left(-\frac{y - y_0}{\alpha}\right) \tag{4.43}$$

The probability of exceedance is defined by

$$p_e(y) = 1 - F_Y(y) = \exp\left(-\frac{y - y_0}{\alpha}\right) \tag{4.44}$$

which can be reorganized into

$$-\log p_e(y) = \frac{y - y_0}{\alpha} \tag{4.45}$$

Therefore, $-\log p_e(y)$ should be a linear function of y, with $\frac{1}{\alpha}$ being the slope and y_0 being the intercept with the abscissa. This forms the basis of distribution fitting to find α and y_0.

How to calculate $-\log p_e(y)$ from the measured ice loads? This can be done through the empirical cumulative distribution function (ECDF) calculated from the event-maxima. If we rank the event-maxima in ascending order, as $Y = \left(Y_1^r, Y_2^r, \ldots Y_N^r\right)$, the ECDF is then

$$\text{ECDF}(y) = \frac{i}{N+1}, \textit{for } Y_i^r < y < Y_{i+1}^r \tag{4.46}$$

We can then plot the minus logarithm of the empirical probability of exceedance, i.e., $-\log(1 - \text{ECDF}(y))$, as a function of y, exemplified in Figure 4.13. By linear fitting to the scatter with minimized least-square error, we can get the distribution parameters α and y_0.

The magnitude of α and y_0 can depend on different parameters, e.g., the instrumented area of ice-load measurement, denoted by A. In the case of ship ramming, ice is assumed to be so thick that it does not break by flexure. Therefore, the thickness of ice does not affect ice loads. Under different sizes of instrumented area (A), we can repeat the distribution fitting process based on measured ice loads in the corresponding area, thereby obtaining different α. This results in the $\alpha - A$ curve as exemplified in Figure 4.14, which is fitted by a power function to get

$$\alpha = 1.25A^{-0.7} \tag{4.47}$$

Eq. (4.47) is obtained by Jordaan et al. (1993). ISO 19906 (ISO, 2019) recommended to use $\alpha = A^{-0.7}$ as the design curve, which is less conservative than Jordaan's initial results.

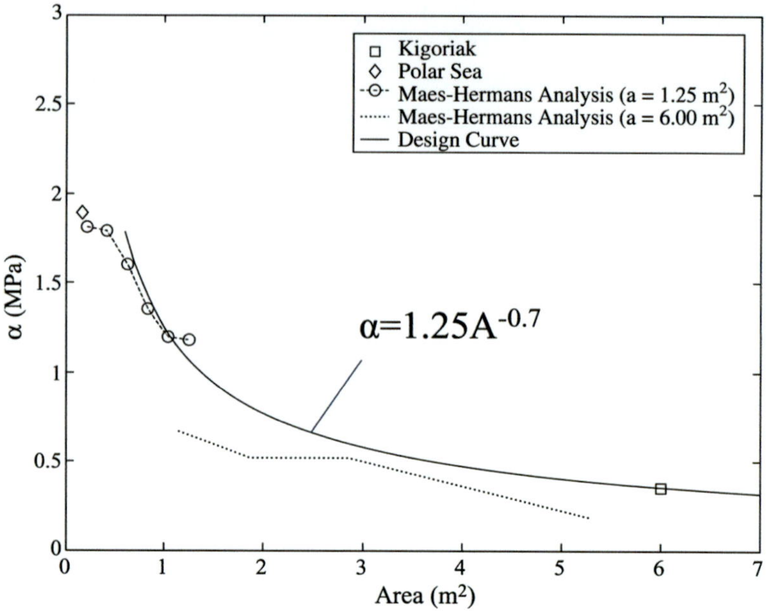

FIGURE 4.14 The α-Area curves.

(Re-produced from Jordaan et al., 1993.)

Now that α can be calculated based on the design area, we can then make long-term prediction of ice loads for the design of structures. If we denote ν as the average number of ramming a ship makes through a year, the number of loading events a ship will encounter through n years is then $n\nu$. With Eq. (4.42) as the parent distribution and follow Eq. (4.39) to derive the extreme value distribution, the CDF of the maximum ice load during this period can be proved to be

$$F_Z(z) = \exp\left(-\exp\left(-\frac{z - y_0 - y_1}{\alpha}\right)\right) \tag{4.48}$$

which is the Gumbel distribution with $y_1 = \alpha(\ln\nu + \ln n)$. If we set $n = 1$, Eq. (4.48) then becomes the CDF of annual maximum load, with $y_1 = \alpha\ln\nu$.

For design, we are often interested in the expected load corresponding to a certain return period, e.g., 100 years. This can be obtained by reorganizing Eq. (4.48) into

$$z_e = y_0 + \alpha\{-\ln[-\ln F_Z(z_e)] + \ln r\} \tag{4.49}$$

where z_e denotes the load level corresponding to the return period. For example, if we are interested in the load at the return period of 100, $F_Z(z_e)$ is then 1–1/100. Jordaan et al. made a conservative assumption that y_0 equals 0. We can then calculate z_e with α depending on the design area.

For example, if we consider a load patch on the hull of a ship, with area being 1 m^2. According to Eq. (4.47), α should be 1.25 MPa. Assume that the ship makes 50 ramming operation in thick ice, we can plot the load level as a function of the

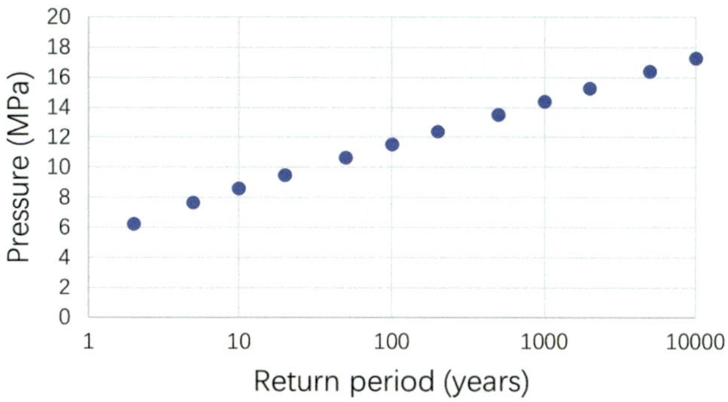

FIGURE 4.15 Ice pressure as a function of the return period.

return period following Eq. (4.41), as illustrated in Figure 4.15. As can be seen, longer return periods correspond to higher ice pressure.

4.3.2.2 Further Development of the Event-Maximum Method

The Event-Maximum Method as introduced above has been included in ISO 19906 standard for Arctic offshore structures. The method has been developed by other researchers during recent years. The development has been carried out in the following aspects:

1. Extension to continuous icebreaking. The original Jordaan et al. method applies for ramming operation in thick ice. Taylor et al. (2010) extended this to continuous icebreaking by adopting an up-crossing methodology to extract load events and maxima from continuous load signals.
2. Inclusion of ice condition. The original Jordaan et al. method applies for ramming thick ice so that ice thickness has no influence in the load level. Shamaei et al. (2020) and later Li et al. (2021) further extended this method to include ice thickness and concentration as the input parameters, which are important factors when ships continuously break ice by bending failure.
3. Modelling of y_0. The original Jordaan et al. method assumes y_0 to be zero. This was shown to be conservative by Taylor et al. (2010) for small design area. Li et al. (2021) proposed a method to model y_0 explicitly as a function of ice thickness and concentration.

In this book, we will not go into the details of these developments. Readers interested in this topic are referred to the references mentioned above.

4.3.3 Modelling of Distribution Parameters to Link Them with Prevailing Ice Conditions

The main restriction of Jordaan et al. approach is the lack of consideration for ice condition parameters. The long-term ice-load estimation is therefore independent of the prevailing ice conditions, such as ice thickness. This can be reasonable in case of vertical structures crushed by ice, or ships ramming very thick ice which does not fail through bending, but can be problematic if the ship mainly operate in continuous mode, breaking ice extensively through bending. Attempts have been made to link probabilistic distribution parameters to ice condition parameters. One of them is the work by Li et al. (2021) which was also mentioned in Section 4.2.2. In this model, the distribution parameter of the line loads, denoted by β, is shown to be dependent on ice thickness and concentration. This allows the estimation of extreme ice loads based on the full operational profile of a ship, accounting for the ice conditions it may encounter and the corresponding operational time in each ice condition.

Kujala (1996) adopt the extreme value theory to establish a semi-empirical probabilistic method of ice loads. In this method, 12-hour load maxima are fitted by the Gumbel distribution, which is a type of extreme value distribution, as per

$$f_Y(y) = \frac{1}{\sigma} \exp\left(-\frac{y-\mu}{\sigma}\right) \exp\left(-\exp\left(-\frac{y-\mu}{\sigma}\right)\right)$$

The distribution parameters, μ and σ, are linked to ice thickness through the mean load maxima via a bi-linear regression function, which enables long-term prediction of ice loads. Kujala method is unique because the author proposed a technique to account for the effect of ship hull angle on the distribution parameters, which enables generalization of the developed model to be used on other ships.

4.3.4 High-Pressure Zone (HPZ) Method

The methods introduced above are essentially *data-driven*, where an ice-load dataset with corresponding ice condition measurement should be present as the basis of modelling. An alternative methodology to this is to derive the ice-load distribution through theoretical ice-load estimation methods, based on the stochasticity in ice conditions. We call this the *mechanics-driven approach*, since the randomness in ice loads originates from the randomness in ice condition, and is derived from the mechanics of ship-ice interaction.

A representative example of mechanics-driven approach is the High-Pressure Zone (HPZ) method proposed by Kujala (1994). This method is established based on the fact that the contact force between a ship and ice is exerted through discrete HPZs, as illustrated in Figure 4.16(a). The number of HPZs (denoted by N) and the magnitude of the total HPZ force (denoted by Y) on a given nominal contact area (denoted by A) are random variables, which can be modelled by Poisson and Gamma distribution, denoted by $f_{N|A}$ and $f_{Y|N}$, respectively. The contact area grows as the ship indents into the ice sheet and is eventually limited by the bending failure, which is determined by ice thickness and strength. Assuming the ice thickness and strength follow Gamma distribution, we can then calculate the probability distribution of the contact area, denoted by f_A, using a theoretical method and then calculate the probability distribution of the total ice load as:

$$f_Y = f_A f_{N|A} f_{Y|N} \tag{4.50}$$

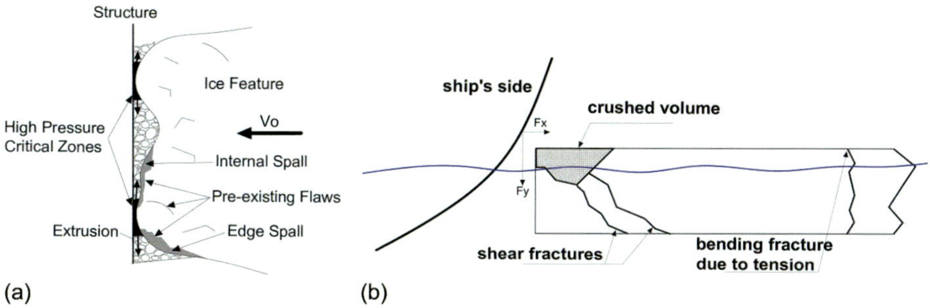

FIGURE 4.16 Interaction between ice and structure: (a) High-Pressure Zones on the contact (Jordaan, 2001), and (b) ice force limited by bending and shearing failure.

(Re-produced from Daley et al., 1998.)

Readers interested in this topic are referred to Kujala (1994) for more details. The original HPZ method was proposed for continuous icebreaking. Later, Ralph (2016) applies it to ship ramming thick ice and improved the method to include speed effect.

Other probabilistic methods existing in the literature include Kotilainen et al. (2017), Suyuthi et al. (2013), Chai et al. (2020), Sinsabvarodom et al. (2020), etc. Readers who are interested in this topic are referred to these articles for further reading.

4.4 SUMMARY

There are multiple ways to calculate ice loads on ships and offshore structures. Popov's energy-based method has been used for decades and become the basis of international ice-class rules. Nevel's narrow wedge theory reveals the important parameter of characteristic length, and provides a closed-form solution for ice bending under static loading. Probabilistic methods developed on full-scale measurements account for the random nature of ice loading and offers insights of long-term ice loads. Besides the methods reviewed here, in the real engineering applications, empirical formulae established on experience are also widely used. Different methods have their pros and cons. Selection of the proper ice-load calculation method hence depends on the target of use and availability of resources.

REFERENCES

Chai, W., Leira, B. J., Naess, A., Høyland, K., and Ehlers, S., 2020. Development of environmental contours for first-year ice ridge statistics. *Structural Safety*, 87, 101996.

Daley, C., 1999, May. Energy based ice collision forces. In *Proceedings of the 15th International Conference on Port and Ocean Engineering under Arctic Conditions (POAC)*, Vol. 2, pp. 674–686, Espoo, Finland, August 23–27, 1999.

Daley, C., Tuhkuri, J., and Riska, K., 1998. The role of discrete failures in local ice loads. *Cold Regions Science and Technology*, 27(3), 197–211.

IACS, 2016. *Requirements concerning polar class*. London: International Association of Classification Societies.

ISO, 2019. Petroleum and natural gas industries — Arctic offshore structures. ISO 19906. https://www.iso.org/standard/65477.html [18.03.2026].

Jiang, Z., 2024. *Evaluating hydrodynamic influences on ice load from ship-ice glancing impact: The roles of sea waves and hydrodynamic interactions*. Doctor thesis, Aalto University, School of Engineering.

Jordaan, I. J., 2001. Mechanics of ice–structure interaction. *Engineering Fracture Mechanics*, 68, 1923–1960.

Jordaan, I. J., Maes, M. A., Brown, P. W., and Hermans, I. P., 1993. Probabilistic analysis of local ice pressures. *The Journal of Offshore Mechanics and Arctic Engineering*, 115(1), 83–89.

Kotilainen, M., Vanhatalo, J., Suominen, M., and Kujala, P., 2017. Predicting ice-induced load amplitudes on ship bow conditional on ice thickness and ship speed in the Baltic Sea. *Cold Regions Science and Technology*, 135, 116–126.

Kujala, P., 1994. *On the statistics of ice loads on ship hull in the Baltic*. Doctoral thesis. Espoo: Helsinki University of Technology.

Kujala P., 1996. Semi-empirical evaluation of long term ice loads on a ship hull. *Marine Structures*, 9(9), 849–871.

Li, F., Suominen, M., Lu, L., Kujala, P., and Taylor, R., 2021. A probabilistic method for long-term estimation of ice loads on ship hull. *Structural Safety*, 93, 102130.

Nevel, D. E. (1961). *The narrow free infinite wedge on an elastic foundation*. U.S. Army Cold Regions Research and Engineering Laboratory, Corps of Engineers.

Popov, Y., Faddeyev, O., Kheisin, D., and Yalovlev, A., 1967. *Strength of ships sailing in ice*. US Army Foreign Science and Technology Center (Translation).

Ralph, F. E., 2016. *Design of ships and offshore structures: A probabilistic approach for multi-year ice and iceberg impact loads for decision-making with uncertainty*. Doctoral dissertation. Memorial University of Newfoundland.

Riska, K., 1987. On the mechanics of the ramming interaction between a ship and a massive ice floe. *VTT Publications*, 43, 86.

Shamaei, F., Bergström, M., Li, F., Taylor, R., and Kujala, P., 2020. Local pressures for ships in ice: Probabilistic analysis of full-scale line-load data. *Marine Structures*, 74, 102822.

Sinsabvarodom, C., Chai, W., Leira, B. J., Høyland, K. V., and Naess, A., 2020. Uncertainty assessments of structural loading due to first year ice based on the ISO standard by using Monte-Carlo simulation. *Ocean Engineering*, 198, 106935.

Suominen, M., Kujala, P., Romanoff, J., and Remes, H., 2017. Influence of load length on short-term ice load statistics in full-scale. *Marine Structures*, 52, 153–172.

Suyuthi, A., Leira, B. J., and Riska, K., 2013. Statistics of local ice load peaks on ship hulls. *Structural Safety*, 40, 1–10.

Taylor, R. S., Jordaan, I. J., Li, C., and Sudom, D., 2010. Local design pressures for structures in ice: Analysis of full-scale data. *The Journal of Offshore Mechanics and Arctic Engineering*, 132(3), 031502.

5 Ship Performance in Ice

5.1 RESISTANCE IN ICE

5.1.1 Background for Resistance in Ice

Ship encounters different type of ice conditions during the operations in sea ice conditions that vary from typical first-year ice conditions to multiyear ice conditions. Although a ship may encounter different types of ice, the operational performance is commonly defined only in some dedicated ice conditions, namely level ice, ice channel and ridged ice. The practical reason for this is that the other ice conditions may be too difficult to define and test ambiguously, like broken ice field, especially in full scale, or the ship should avoid the ice type and the encounter could lead to severe consequences, like encountering an iceberg.

From the mentioned ice conditions, level ice does exist in nature on a scale that we could say that a ship is operating under level ice conditions. However, it is a basic assumption that gives us a possibility to compare the icebreaking capability and performance of vessels. Furthermore, full-scale measurements involve several uncertainties that justify accepting relatively major ice thickness deviations during tests, and model-scale tests provide an opportunity to compare the level ice performance of different designs. In addition, level ice is also a common assumption for the development of theoretical models.

Merchant vessels are commonly designed to operate in icebreaker assistance. Thus, performance in a newly broken channel or in an old brash ice channel is important design criteria for merchant vessels that are not designed for independent operations. Generally, a brash ice channel is an easier ice condition compared to level ice as the ship does not need to break the ice that significantly reduces the resistance imposed by the encountered ice.

Ridged ice is the most severe ice condition that icebreakers are typically designed to encounter during common operations. Merchant vessels are not commonly designed to operate in ridged ice field independently. As mentioned above, ships may encounter ice conditions that are more severe than ridged ice, like icebergs, but these encounters are not considered as common operations. In case a ship encounters heavy ridges, the propulsion is not sufficient to overcome the resistance imposed by the ridges, and the ship penetrates the ridge with the help of inertia and by ramming the ridge multiple times.

5.1.2 Formulation of Resistance in Ice

The first resistance formulations for ships in ice were developed in 1889 by Runeberg. After the pioneering work by Runeberg (1889), there was a prolonged period during which no significant advances were made in research and theoretical work related to ship performance in ice. The work mainly focused on practical solutions for ships

DOI: 10.1201/9781003676829-5

operating in ice. In the 1960s and 1970s, the research related to ship resistance in ice gained new interest and significant development was made, especially by Kashteljan and Enkvist. The research focused mainly on level ice conditions and several formulations for ship resistance in ice were developed by different authors. This development was followed by studies related to ship resistance in ridged ice field by Keinonen in the late 1970s, and ship resistance in brash ice channel by Mellor and Malmberg in the early 1980s. These studies were later followed by other researchers and new knowledge and formulations have been developed.

Kashteljan (1968) was the first to divide the ice resistance into components that has been the basis for majority of the developed formulations ever since. The principal idea of this approach is that the ice resistance is composed of resistance components associated with different processes or phenomena that can be combined according to the superposition principle. It should be noted that this approach assumes the processes are not coupled, i.e., not dependent on other processes. It is assumed that the total resistance, R_{tot}, is composed of resistance in ice, R_{ice}, and open water, R_{ow},

$$R_{tot} = R_{ice} + R_{ow} \tag{5.1}$$

The resistance in ice can be further divided into components related to different processes and forces resisting the ship movements (see Figures 5.1 and 5.2). The division of components may differ significantly between the sources and authors

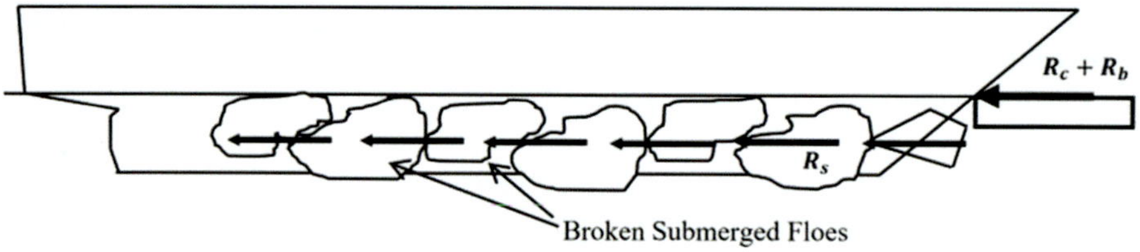

FIGURE 5.1 Resistance components related crushing, R_c, bending, R_b, and submerging, R_s, of ice acting on different hull regions (Riska et al., 2001).

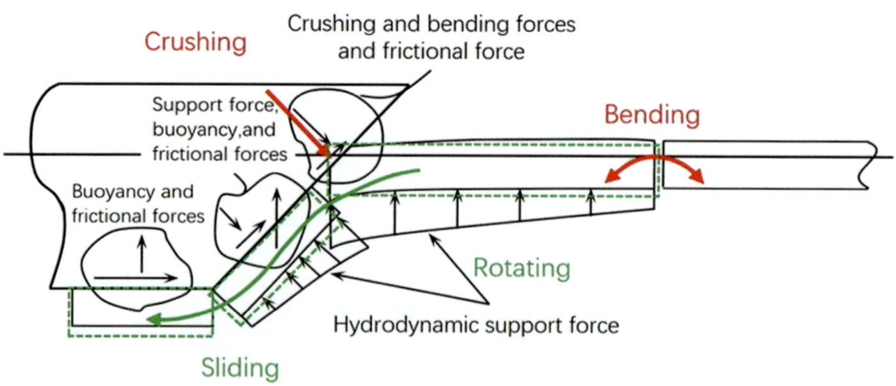

FIGURE 5.2 Forces related to breaking (Riska, 2017).

depending on their assumptions and focus areas. In general, it can be claimed that the main resistance components in ice can be divided into components related to breaking the ice and submerging and displacing the ice, while the speed of the vessel and friction coefficient between the hull and ice are accounted in these components or into separate components. The breaking component can be further divided into crushing and bending as depicted in Figures 5.1 and 5.2.

The resistance components are then affected by different ice, ship and operational parameters. Typically, the formulations account for the thickness and strength of ice, the friction coefficient between the ship hull and ice, and density of ice from ice-related parameters. Ship-related parameters affecting the resistance include the main dimensions and hull form of the vessel, while the operational parameters account for the speed of the vessel.

The developed equations are commonly based on either full-scale and or model-scale measurements and determined by regression analysis. The parameters affecting each resistance component are determined from this analysis. Typically, the resistance components, R_{comp}, are of the form:

$$R_{comp} = C*L^x*B^y*\sigma_f^z*h_i^u*v^v*\mu^w \tag{5.2}$$

where C is a constant, L and B are the length and breadth of the vessel, σ_f and h_i are the strength and thickness of ice, v is the speed of the ship, and μ is the friction coefficient between the hull and ice. It should be noted that the parameters may vary, like the draught or hull scale may be accounted for, or the compressive strength of ice. However, the form is similar. When using this type of formulations, it is important to note the assumptions and the data on which the equations are based before applying them.

The earlier studies have shown that the hull design (shape) and main dimensions of the vessel have significant impact on the resistance and may have an impact on the portion of the different resistance components. Small stem and frame angles enhance ice breaking through bending that decreases the resistance, while vertical hull shapes favour breaking through crushing. Similarly, sloping hull forms enhance submerging of ice while vertical angles enhance pushing the broken ice away from the vessel. The breadth of the vessel determines how much ice needs to be broken and displaced, thereby increasing the resistance. Longer vessels have a larger contact area for the broken ice that increases the resistance.

Studies have indicated that ice resistance increase linearly as a function of speed, see e.g. Lindqvist (1989). However, as schematically presented in Figure 5.3, the resistance does not start from zero and resistance at zero speed is unambiguous. The former can be justified by the fact that ice needs to be broken, regardless of the speed, that causes resistance to the ship. For the later, the ship is idling with zero speed in ice condition and thrust is increased in steps. When the thrust reaches the force required to break the ice, point A in Figure 5.3, the ship starts to move. When the thrust is increased, the speed increases linearly as a function of added thrust, from point C to D in Figure 5.3. However, when the thrust is decreased, the ship is able to move with thrust values that are lower than the level that was required to make the ship move, from point C to B in Figure 5.3. This has been observed in full-scale and model-scale tests and is considered to be related to static and dynamic friction.

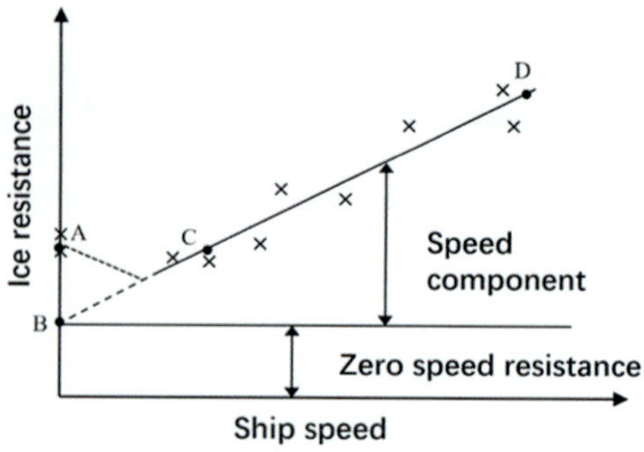

FIGURE 5.3 Schematic presentation about the influence of the speed on ice resistance.

(Reproduced form Kujala and Riska 2010.)

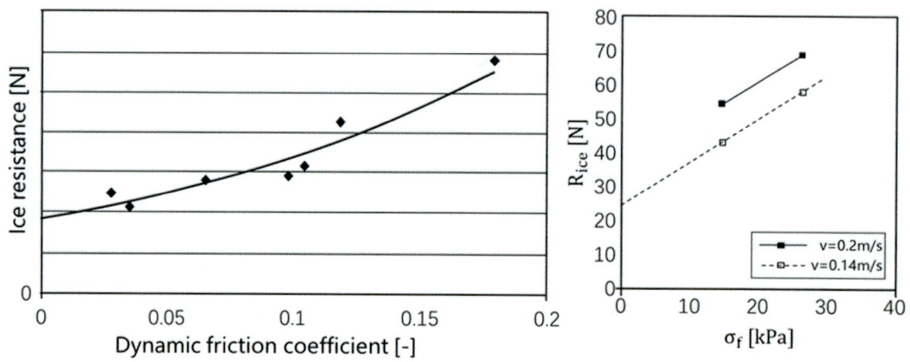

FIGURE 5.4 Ice resistance as a function of dynamic friction coefficient and flexural strength on the left and right, respectively.

(Reproduced from Enkvist and Mustamaki, 1986.)

From the ice conditions, the thickness has the most significant impact to the resistance. The impact varies between the studies, but ice resistance has been shown to be proportional to ice thickness to power 1.0–2.2 (Myland, 2019). Besides ice thickness, the dynamic friction coefficient and flexural strength of ice have a significant impact on the resistance as indicated in Figure 5.4.

5.1.3 Resistance in Level Ice

As mentioned above, several studies have focused on level ice resistance and defined their own formulation to estimate the resistance. Depending on the assumptions made and the available data, these formulations may deviate from the division and share of resistance components, accounted for ship and ice parameters, etc. This chapter presents Lindqvist's (1989) approach as an example as it has been shown to give a good indication of the level ice resistance for conventional icebreaking hull forms by many independent studies, see e.g. Myland (2019). This presentation is a summary of the derivation. More accurate description is provided in Lindqvist (1989).

Lindqvist divided the resistance in three main components: crushing at stem R_c, bending failure at bow R_b, and submerging R_s, and speed effect. The maximum vertical force, F_v, at stem was related to flexural strength of ice, σ_f, and ice thickness, h_i, as follows:

$$F_v = 0.5\sigma_f {h_i}^2 \tag{5.3}$$

Analysing the force components using cumulative friction and assuming that the friction force acts along the verticals, the crushing-related resistance at stem was derived to have a form

$$R_c = 0.5\sigma_f {h_i}^2 \frac{\left(\tan\varphi + \dfrac{\mu_H \cos\varphi}{\cos\psi}\right)}{\left(1 - \dfrac{\mu_H \sin\varphi}{\cos\psi}\right)} \tag{5.4}$$

where μ_H is the friction coefficient and the hull angles, φ and ψ, are depicted in Figure 5.5. For the bending component, the shape of the ice edge was assumed to be triangular, see Figure 5.6, and the vertical component was defined based on the basic strength of materials theory. To account for the bending forces at the whole bow area, the size of the broken pieces was defined that was utilized to determine the number of simultaneous local events. The total vertical force was then defined based on the number of events. Using the above assumptions and geometrical relationships, the vertical force was then converted to the horizontal resistance components as follow:

$$R_b = 0.003\sigma f B(h_i)^{3/2}\left(\tan\psi + \frac{\mu_H \cos\varphi}{\sin\alpha\cos\psi}\right)\left(1 + \frac{1}{\cos\psi}\right) \tag{5.5}$$

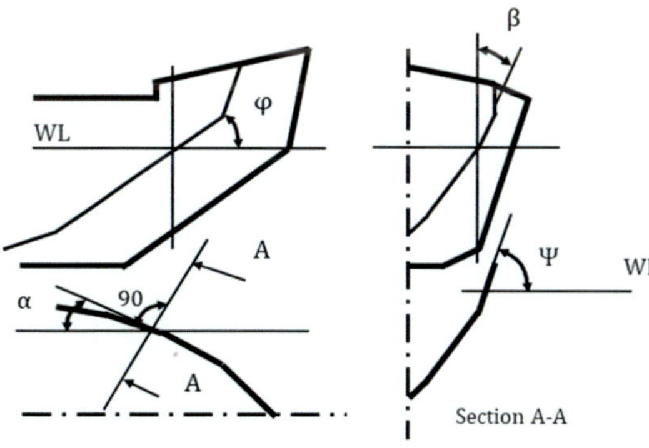

FIGURE 5.5 Notation related to hull angles.

(Reproduced from Kujala and Riska, 2010.)

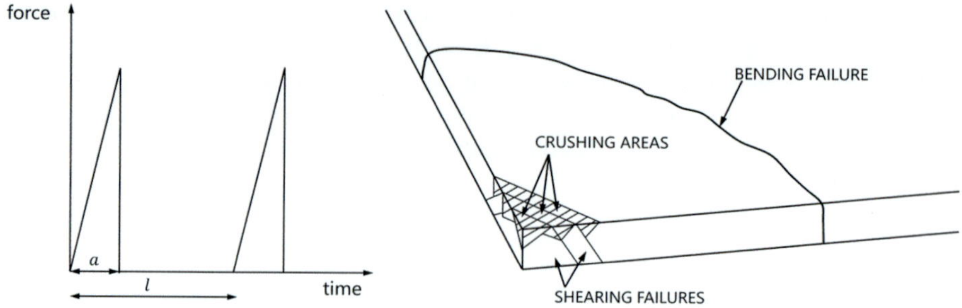

FIGURE 5.6 Schematic presentation of breaking by bending.

(Re-produced from Lindqvist, 1989.)

Submerging-related resistance was considered to consist of potential energy required to submerge the ice and frictional-related resistance. For the former, the breadth of the vessel, B, defines the extent of the submerged ice. Here the ice under the keel is submerged to a distance equal to the draught of the vessel, T, and the ice at the side of the vessel is submerged to a distance equal to T/2. For the later part, it is considered that the hydrostatic pressure is pressing the ice against the hull causing friction-related resistance. Here, it was assumed that 70% of the bottom is covered by ice, and the pieces moving along the ship bow are causing frictional forces. Following these assumptions the component for the submerging-related resistance was derived:

$$R_s = (\rho_w - \rho_i) g h_i B \left[\frac{T(B+T)}{B+2T} + \mu_H \left(0.7L - \frac{T}{\tan\varphi} - \frac{B}{4\tan\alpha} + T\cos\varphi\cos\psi\sqrt{\frac{1}{\sin\varphi^2} + \frac{1}{\tan\alpha^2}} \right) \right] \quad (5.6)$$

Speed, v, was assumed to affect all the components, and accounted by multiplying all the components with a speed correction factor. The final equation has the form:

$$R_{ice} = (R_c + R_b)\left(1 + 1.4\frac{v}{\sqrt{gh_i}}\right) + R_s\left(1 + 9.4\frac{v}{\sqrt{gL}}\right) \quad (5.7)$$

where g is the gravitational acceleration, and L is the length of the vessel.

5.1.4 Resistance in Ridged Ice and Brash Ice Channel

Although several formulations for the level ice resistance have been formulated, not so many formulations exist for ridged ice and brash ice channel. In these cases, forming a material model for ridged and brash ice is complicated as the material consist of solid ice pieces that are loosely connected. The main ice-related parameters are porosity, piece size and thickness. However, in case the piece size is clearly smaller than the total thickness, soil mechanics can be applied to form the material model. This can be considered acceptable assumption for large ridges and thick old brash ice channels, although the piece size is relatively larger when compared to traditional soils like clay. This approach has been utilized in the research that has formed the base for the resistance calculation applied in Finnish-Swedish Ice-Class Rules. However, other approaches have also been developed, and the presentation here should be considered as an example on how the resistance in a ridged ice and brash ice channel can be formulated.

The main parameters of the material models of soil mechanics are internal friction, related to friction angle ϕ and cohesion, c. The internal friction describes the internal forces related to moving the particles with respect to each other, while cohesion describes the force acting between the particles that try to prevent deformation. In these approaches, the material is considered to be continuous. Mohr-Coulomb material model offers a possibility to model shear force that accounts for cohesion and internal friction. However, it leads to complicated equations. Applying Rankine's model, that assumes cohesion to be zero, simplifies the equations significantly (Mellor, 1980). This can be considered a valid assumption, as in nature the brash ice channel and ridge keel consist of solid ice blocks that are loosely connected to each other and the cohesion is commonly negligible, but not actually zero (Matala and Suominen, 2023).

Figure 5.7 represents the formation of slip planes in the mass that follows Rankine's material model when it is pushed by a vertical wall. Here, the slip lines

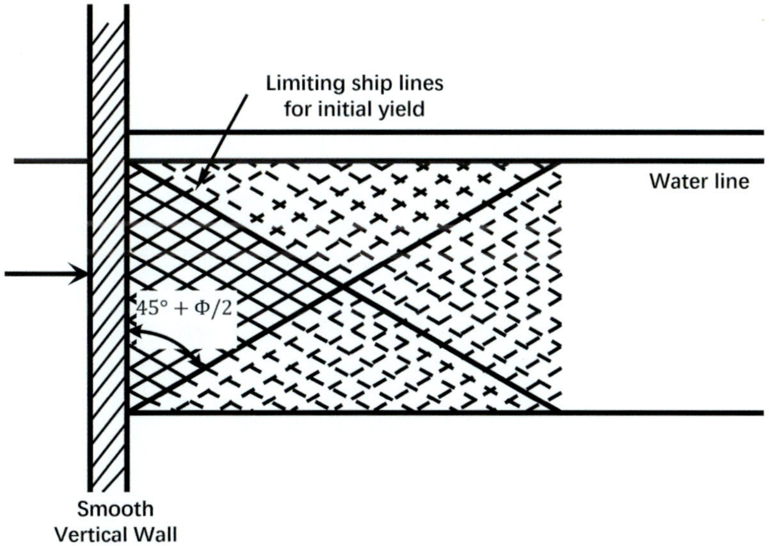

FIGURE 5.7 A principal picture from slip planes when cohesionless material is pushed.

(Reproduced from Mellor, 1980.)

form an angle of 45°+ ϕ/2 with respect to the vertical wall. The friction angle is a material constant that depends on the friction between the particles and other forces resisting the shear deformation (Kujala and Riska, 2010).

When the mass is at rest and gravity is the only body force acting on it, the vertical stress, σ_z, can be calculated with an equation

$$\sigma_z = (1 - p)(\rho_w - \rho_i)gz \tag{5.8}$$

where p is porosity, ρ_w and ρ_i are density of water and ice, respectively, and z is the vertical distance from the keel. The maximum stress occurs at waterline, where the ice above the water presses the ice mass downward due to gravity with the same force the buoyancy is lifting the ice. Assuming plane stress situation, the horizontal stress can be described as:

$$\sigma_x = \sigma_y = \frac{\nu}{1-\nu}\sigma = K_0\sigma_z \tag{5.9}$$

where ν is Poisson ratio and K_0 is pressure factor related to mass at rest. In case the mass is affected by forces, it is no longer in rest but in passive or active stage. The passive and active stages define the maximum stress before the material yields and forms slip planes in compression and tension, respectively.

$$\sigma_x = K_p\sigma_z = \left(\frac{1 + \sin\varphi}{1 - \sin\varphi}\right)\sigma_z \tag{5.10}$$

$$\sigma_x = K_a\sigma_z = \left(\frac{1 - \sin\varphi}{1 + \sin\varphi}\right)\sigma_z \tag{5.11}$$

Here, K_p and K_a denote the passive and active stress factors, respectively. These stress factors are presented as a function of internal friction angle in Figure 5.8. It should be noted that the stress factors are sensitive to the friction angle. Thus, accurate measurements are important.

Noting the shape of the channel, we can assume a passive stage occurs in the middle and active at the edges. Denoting channel thickness at the edges and middle with H_1 and H_2, the total force at the edge, F_a, and in the middle, F_b, can be denoted as

$$F_a = \frac{1}{2}\sigma_y H_1 = K_a(1 - p)(\rho_w - \rho_i)gH_1^2 \tag{5.12}$$

$$F_b = \frac{1}{2}\sigma_y H_2 = K_b(1 - p)(\rho_w - \rho_i)gH_2^2 \tag{5.13}$$

It follows that

$$\frac{H_1}{H_2} = \sqrt{\frac{K_p}{K_a}} = \frac{1 + \sin\varphi}{1 - \sin\varphi} \tag{5.14}$$

It typically has values ranging from 5 to 12 with the common internal friction angles present in the nature 42° to 58° (Kujala and Riska, 2010).

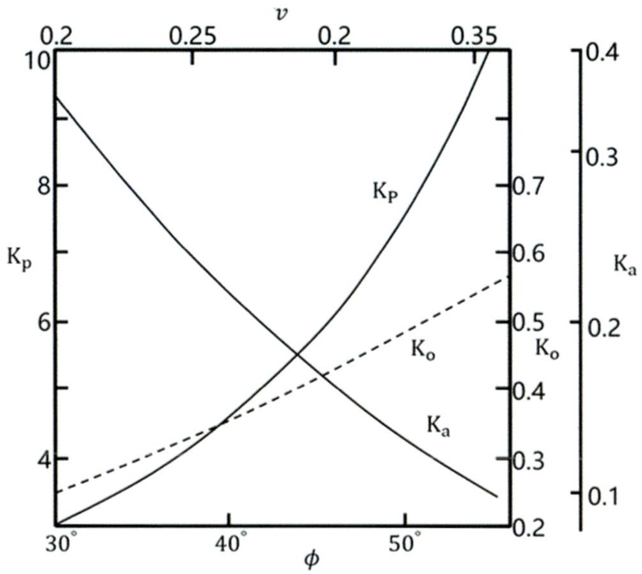

FIGURE 5.8 Stress-related factors K_0, K_a and K_p, as a function of internal friction angle. **(Reproduced from Mellor, 1980.)**

Resistance in ridged ice and brash ice channel can be derived utilizing this material model following the approach described by Malmberg (1983). Figure 5.9 presents a schematic presentation for the derivation of the equation. It should be noted that the resistance is considered to consist of resistance related to displacing the ice at bow, and frictional-related resistance at the bow, midship and bottom of the ship.

Assuming the speed of the ship does not affect the ridge mass, the consolidated layer is negligible, and the ridge mass is at passive stage from a width of $B + 2b$, as depicted in Figure 5.9; the total normal force, F_N, at the side of the bow can be expressed with equation

$$F_N = \frac{1}{2} K_p (1 - p)(\rho_w - \rho_i) gHT(B + 2b) \tag{5.15}$$

Noting the geometry, the force resisting the movement of the ship at bow, F_{x1}, is

$$F_{x1} = F_N (\mu_H \cos\phi + \sin\psi \sin\alpha) \tag{5.16}$$

At midship, the ridge mass is at rest and resists the movement through friction, F_{x2},

$$F_{x2} = \frac{1}{2} \mu_H K_0 (1 - p)(\rho_w - \rho_i) g 2 L_{par} HT \tag{5.17}$$

Assuming the whole keel is covered with ice, the frictional resistance related to buoyancy of submerged ice pieces, F_B, can be expressed as:

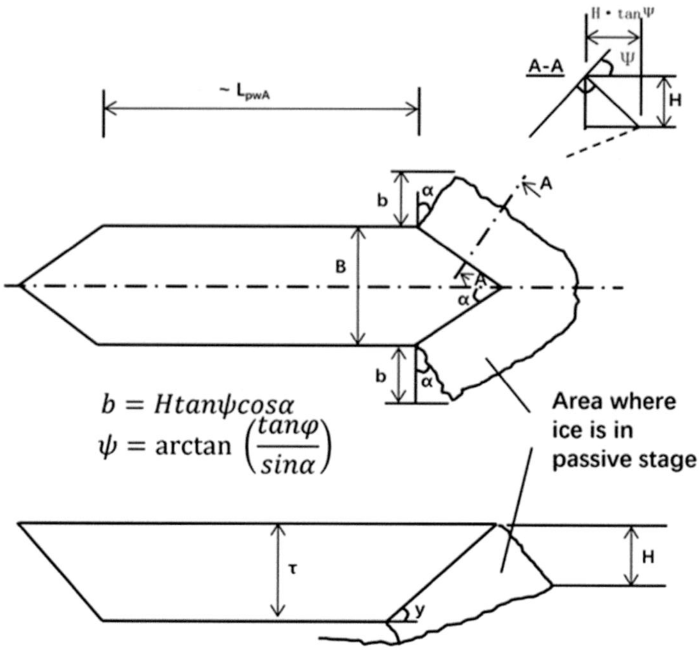

FIGURE 5.9 Schematic presentation from ship penetrating a ridge that is utilized to derive the equation for the resistance in ice.

(Reproduced from Malmberg, 1983.)

$$F_B = \mu_H(1-p)(\rho_w - \rho_i)g\left(H - \frac{1}{2}T\right)L_{par}B \tag{5.18}$$

The resistance related to penetrating the ridge is summation of the above force terms F_{x1}, F_{x2}, and F_B

$$R_R = (1-p)(\rho_w - \rho_i)gT\left\{K_pH\left(\frac{1}{2}B + H\tan\psi\cos\alpha\right)(\mu_H\cos\varphi + \sin\psi\sin\alpha) + \mu_H L_{par}\left(K_0H + \left(\frac{H}{T} - \frac{1}{2}\right)B\right)\right\} \tag{5.19}$$

This formulation has been shown to give a good correspondence with full-scale observation (Malmberg, 1983).

The channel resistance can be derived following the same approach. Denoting channel ice thickness with H and noting the relation $H_1/H_2 = (K_p/K_a)^{1/2}$ the average thickness, H_{average}, can be described as:

$$H_{\text{average}} = \frac{1}{2}H\left(1 + \sqrt{\frac{K_a}{K_p}}\right) \tag{5.20}$$

Assuming the channel thickness is smaller than the draught, the channel resistance, R_{ch}, can be derived to have the form (Malmberg, 1983)

$$R_{ch} = (1-p)(\rho_w - \rho_i)gH^2\left\{\left[\frac{B}{2}\left[\frac{1}{2}\left(1 + \frac{K_a}{K_p}\right)\right]^2 + H\tan\psi\cos\alpha\right]\right.$$

$$\left. K_p(\mu_H\cos\alpha + \sin\psi\sin\alpha) + K_0 L_{par}\mu_H\right\} \tag{5.21}$$

This formulation has been further developed by other authors by adding the effect of speed to the resistance formulation (Riska et al., 1997)

$$R_{ch} = \frac{1}{2}(1-p)(\rho_w - \rho_i)gH_F{}^2K_p\left[\frac{1}{2} + \frac{H_M}{2H_F}\right]^2\left[B + 2H_F\left(\cos\delta - \frac{1}{tan\,\psi}\right)\right]$$

$$(\mu_H\cos\alpha + \sin\psi\sin\alpha) + (1-p)(\rho_w - \rho_i)gK_0\mu_H L_{par}H_F^2$$
$$+ (\rho_w - \rho_i)g\left[\frac{LT}{B^2}\right]^3 H_M A_{WF} F_n^2 \tag{5.22}$$

Here L_{par} is the length of parallel midship, A_{WF} is the waterline area of the bow, angles δ and γ are related to channel profile as described in Figure 5.10 that typically have values 22.6° and 2°, respectively, at the Bay of Bothnia, F_N is Froude number, H_F and H_M are the thickness of the channel at the sides and middle, respectively, that are related to each other by formulation

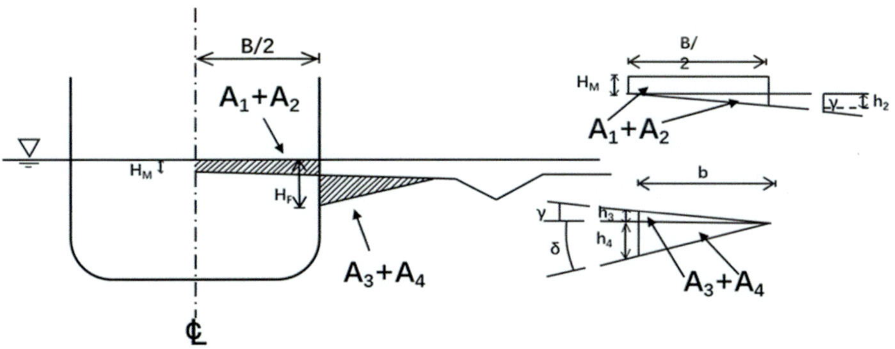

FIGURE 5.10 Determination of the shape of the brash ice at the side of the vessel.

(Reproduced from Riska et al., 1997.)

$$H_F = 0.26 + \sqrt{BH_M} \tag{5.23}$$

where it is assumed that the breadth is larger than ten meters. It should be noted that different types of channels exist in nature and other values could be justified.

The formulation presented here has formed the basis for the brash ice channel resistance calculation that is currently utilized in Finnish-Swedish Ice-Class Rules. It should be noted that the equations have a different format, and different parameters and coefficients are used in the rule calculation. However, the origin is the same in both.

5.2 MANOEUVRING IN ICE

Comparing to manoeuvring in open water, manoeuvrability of ships in ice-covered area is largely constrained by the surrounding ice. Ice introduces additional movement onto the ship, restricting its turning capacity extensively. In level ice fields, the radius of turning circle can be significantly larger than that in open water. In pack ice fields, the constraint is much less as ice are free to move. The turning radius is usually not much different from that in open water.

In addition to ice conditions, manoeuvrability in ice is also influenced by the shape of the ship's hull. Different from operating straight ahead, when turning in ice, the midship and stern shoulder regions of the ship get into heavy contact with ice. Therefore, the form of the stern shoulder and midship is especially important. Ships with long and vertical midship sections typically manoeuvres poorly in ice.

Manoeuvrability of ships is also greatly affected by the option of propulsion. The new azimuth drive with pod propulsion solutions have significantly improved the ship's manoeuvrability in ice, as the ship's thrust can be effectively directed towards different directions. Figure 5.11 presents the turning radius of different ships with various hull forms and propulsion solution. It is obvious that ships with icebreaking hull forms (icebreakers) have much smaller turning radius than others (cargo vessels, tankers), and ships with pod propulsion turns better than those with conventional propulsion solutions (propeller and rudder).

The mathematical model of ship manoeuvring in ice can be set up by slightly extending the corresponding model in open-water conditions. The model can be set up by 3, 4 or 6 degrees of freedom (DoF). Here we present the example with four DoFs, including surge, sway, roll and yaw. The inclusion of roll motion is motivated by the fact that roll motion changes the hull inclination angles, and thus may affect the icebreaking process in the shoulder and midship regions during ship turning. Essentially, the ship performance model solves the following coupled motion equations under ship-fixed coordinate system:

$$\begin{bmatrix} M_{ship} & 0 & 0 & 0 \\ 0 & M_{ship} & 0 & 0 \\ 0 & 0 & I_{xx} & I_{xz} \\ 0 & 0 & I_{zx} & I_{zz} \end{bmatrix} \begin{bmatrix} \dot{v}_x - \omega_z v_y \\ \dot{v}_y + \omega_z v_x \\ \dot{\omega}_x \\ \dot{\omega}_z \end{bmatrix} =$$

$$\sum_i \boldsymbol{F}_{b,i} + \sum_i \boldsymbol{F}_{r,i} + \boldsymbol{R}_s + \boldsymbol{T}_{net} + \boldsymbol{F}_{rud} + \boldsymbol{F}_{hd} + \boldsymbol{F}_{hs} + \boldsymbol{F}_g \tag{5.24}$$

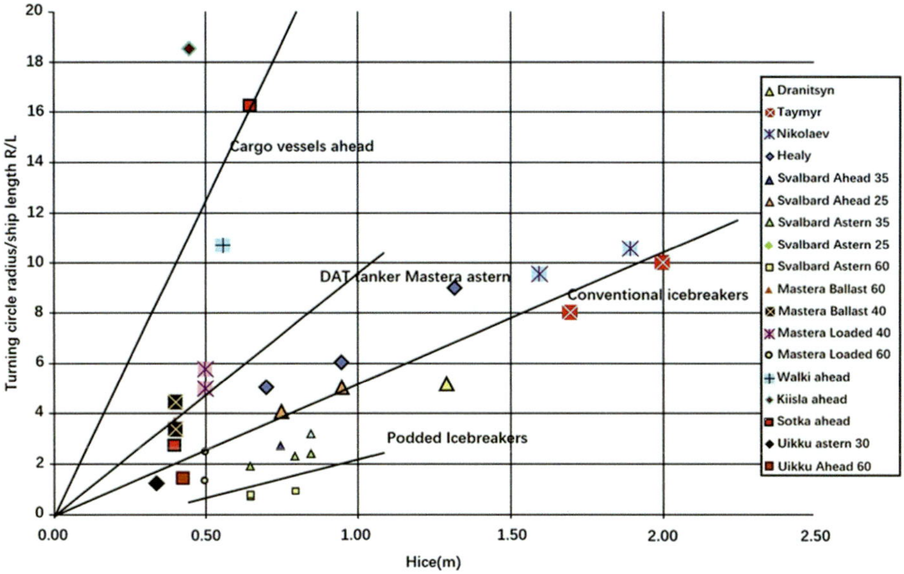

FIGURE 5.11 Turning performance of ships in level ice (Wilkman et al., 2016).

Here $\boldsymbol{F}_{b,i}$, $\boldsymbol{F}_{r,i}$, $\boldsymbol{R}_s$, $\boldsymbol{T}_{net}$, $\boldsymbol{F}_{rud}$, $\boldsymbol{F}_{hd}$, $\boldsymbol{F}_{hs}$, and $\boldsymbol{F}_g$ are the local ice breaking and rotation force the force to turn broken ice against the hull, submersion resistance, propeller net thrust, rudder force, hydrodynamic and hydrostatic force, as well as gravity force. M_{ship} is ship mass and I_{xx}, I_{zz} and I_{xz}, I_{zx} are the moments of inertia. The speed terms are denoted by v_x, v_y, ω_x, and ω_z. Unlike resistance calculation, there has been no direct method to calculate the resisting moment or the turning radius in ice. Currently, numerical simulation is the only way to solve the problem and estimate the manoeuvrability of ships. This topic will be introduced in Chapter 8.

The manoeuvrability of a ship in ice is most commonly measured by the turning circle. This involves not only the size of the turning circle but also the time taken to complete the turn. The turning circle is typically presented relative to the ship's length, as shown in Figure 5.12. In addition to the turning circle, other operations that illustrate manoeuvrability for icebreakers include the star-manoeuvre. Star-manoeuvre is used when the crew intends to reverse the heading of the ship when an icebreaker receives a mission of icebreaking assistance. The crew will manoeuvre the ship by going back and forth with the rudder or pod operated, thereby making its way to reverse. Figure 5.12 illustrates the course of a typical star-manoeuvre, which is also referred to as the captain's turn. The key measure of star-manoeuvring performance is the time required to reverse the heading.

Another typical mode of manoeuvre in ice is to break out of an ice channel. This happens when a ship needs to leave an ice channel which it has been sailing in. The key measure of ice channel break-out capability is the distance required between the starting point of turning manoeuvre and the ending point of completely leaving

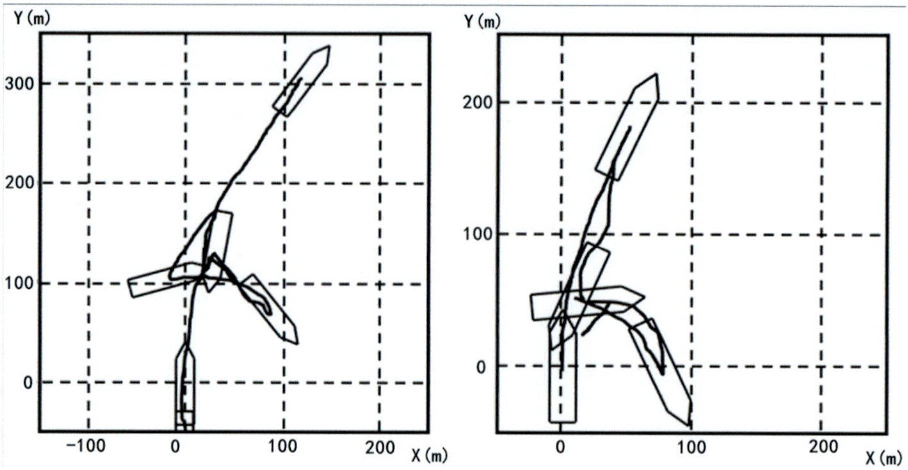

FIGURE 5.12 Typical star-manoeuvre courses.

(Reproduced from Kujala and Riska, 2010.)

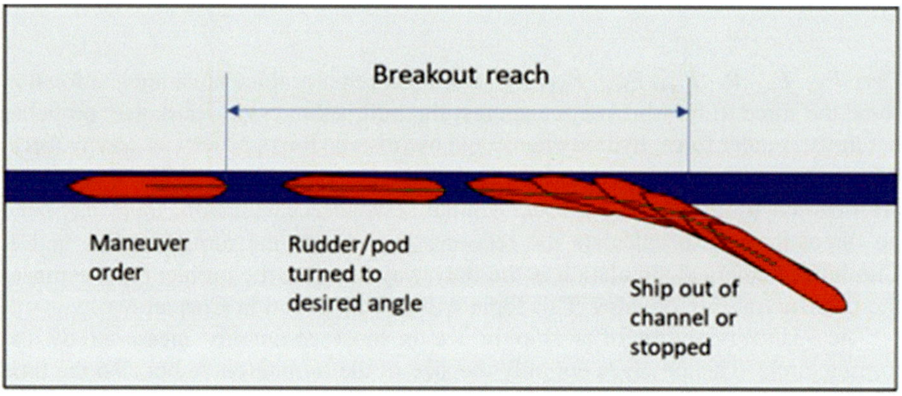

FIGURE 5.13 Illustration of break-out manoeuvre and definition of breakout reach (Li et al., 2024).

the ice channel (see Figure 5.13), termed as the *break-out reach*. The break-out reach depends heavily on the width of the channel in relation to the ship width. In addition, it also depends on the initial speed when turning starts, ship hull form, ice thickness and propulsion solution. The effect of the above factors on break-out reach is discussed by Li et al. (2024) through a set of model-scale tests.

Turning circle test, breaking out of channel test and star manoeuvre are the three standard manoeuvring tests in ice as recommended by ITTC (2024).

5.3 PROPULSION IN ICE

5.3.1 The History of Propulsion in Ice

Wikman et al. (2019) presented an introduction of the history of propulsion in ice. The propulsion of ice-going ships has undergone significant evolvement during the past decades. In the beginning, propulsion solutions were modifications of well-proven technology, with fixed pitch and controllable pitch propellers connected directly or through a gearbox to low/medium speed diesel engines. In some applications, the propeller was equipped with a nozzle. With the need for better manoeuvrability in ice, ship propulsion systems have evolved from conventional direct shaft lines to azimuth devices, which can turn 360 degrees.

One of the early series of ships was the Amguema-class general cargo vessels. These diesel-electric vessels were built between 1962 and 1975 and were pioneers in the Arctic trade. *MV (Motor Vessel) Arctic* is an icebreaking cargo ship built in 1978. The propulsion machinery of the vessel comprises a slow-speed, two-stroke diesel engine connected to a controllable pitch propeller with a diameter of 5.23 m in a nozzle. In the middle of the 1970s, significant advancement in device development took place. For example, the dry cargo ships of the Russian Maritime Register of Shipping UL-class (today LU4) were ordered from East Germany, Warnowwerft. The propulsion plant consisted of one slow-speed diesel engine connected directly to a fixed-pitch propeller. The machinery had no modern electronic control and thus at low speeds the maximum power of the engine (8200 kW) could not be utilized. In addition, the multi-purpose cargo ships, *SA15*, were built in Finland with two medium-speed diesel engines followed by clutches, hydrodynamic couplings, reduction gear, elastic couplings and single controllable pitch propeller.

In the 1980s, the boom to build nuclear ships in the Soviet Union was blooming. The huge barge/container carrier vessel, *Sevmorput*, has one nuclear reactor, and a steam turbine connected to one controllable pitch propeller in a nozzle. The total power was 29400 kW.

The development of propulsion took giant leaps towards azimuth drives in the 1990s. Icebreakers and tankers in the Baltic Sea trade were installed with both mechanical Z-drives and electric-podded drives. The biggest units were installed onboard two 106,000 DWT tankers, which are Tempera and Mastera of Neste shipping, see Figure 5.14. These tankers showed the icebreaking possibilities that had never been seen before; they were the first so-called double-acting tankers, meaning that in heavy ice they proceeded astern to break the ice. This enables optimizing the ship bow for open water performance and astern is designed to break the ice. In practice they also did some icebreaker duties by assisting other merchant ships as shown in Figure 5.14.

Today more azimuth propellers have been used on a variety of icebreaking vessels, with the power applicable depending on the vessel's needs. The first vessel equipped with a 13 MW Azipod VI2300 propulsor, *Norilsk Nickel*, was a highly successful project and it has broken the way to other Arctic projects (see Figure 5.15).

Another interesting and innovative concept enabled by azimuthing propulsion is the so-called oblique icebreaker developed by Aker Arctic in the 2010s (see Figure 5.16).

(a) (b)

FIGURE 5.14 *MT Tempera* Azipod propulsion and ship in operation in the Gulf of Finland.

(Courtesy of ABB.)

(a) (b)

FIGURE 5.15 Azipod installed on the *Norilsk Nickel*.

(Courtesy of ABB.)

It has mechanical azimuthing drives as shown in Figure 5.16. The main idea was that the ship can break ice sideways enabling much wider channel than the traditional icebreakers can do. The hull is unsymmetric which naturally makes the design of the vessel very demanding, especially the seakeeping on waves is an interesting challenge to handle both in-design phase as well as operation phase.

The latest development has been the so-called Yamal LNG project enabling year-round independent navigation in the Northern Sea Route (see Figure 5.17). The ship has ice-class PC 2 and she can break more than 2 m thick ice.

Finland has a long history of building icebreakers and Figure 5.18 gives a summary of the used propulsion systems on those. The first ever diesel-electric propulsion system was used on *IB Sisu* built in Helsinki shipyard having two shaft lines aft and one in front, with a power of 1335 hp each. *IB Voima* was built in 1954 in Helsinki shipyard, which have two shaft lines aft and two in front, with a power of

(a) (b)

FIGURE 5.16 Oblique icebreaker *Baltika* developed by Aker Arctic and used propulsion system.

(Courtesy of Aker Arctic and Steerprop.)

2574 kW each. *IB Urho* had also four shaft lines built in 1973 in Helsinki shipyard, having the power of 2455 kW each at aft and 1605 kW at front. *IB Otso* was built in 1986 in Helsinki, having only two shaft lines aft, with a power of 7500 kW each and an air-bubbling propulsion in front with a power of 1900 kW. *IB Fennica* was the first multipurpose icebreaker built in Aker-Finnyards in Rauma in 1993; she was the first icebreaker having azimuthing propulsion systems, two mechanical systems at aft with a power of 7500 kW each (see also Figure 5.19). *IB Botnica* was built in 1998 in Aker-Finnyards equipped with two azimuthing propulsion systems, now in Azipod style, with a power of 5 MW each. The newest Finnish icebreaker is *Polaris* built in 2016 in Helsinki shipyard, equipped with three azimuthing propulsion systems, two aft and one in front, with a power of 6500 kW aft each and 6000 kW in front (see also Figure 5.20).

5.3.2 Propulsion Efficiency in Ice

Determination of propulsion coefficients largely depends on test measurements in both model and full scale. The determination of these coefficients in open water has a long history, and the testing methods are generally accepted and widely used. However, in ice conditions, these coefficients have been studied less extensively. The presence of ice and the propeller's load near bollard pull conditions complicate the reliable determination of these coefficients in many ways. The following is a brief review of the literature in this field.

5.3.2.1 Icebreaker Propeller Design (Enkvist and Johansson, 1968)

Enkvist and Johansson deal with the propeller design of an icebreaker from several perspectives. They discuss the effect of ice blocks on propeller performance and required strength, the importance of propeller diameter and pitch ratio, and the effect

(a)

(b)

FIGURE 5.17 Three gigantic 15 MW Azipod units installed on *LNGC Christophe De Margerie*. The vessel dimensions are length 299 m and breadth 50 m.

(Courtesy of ABB.)

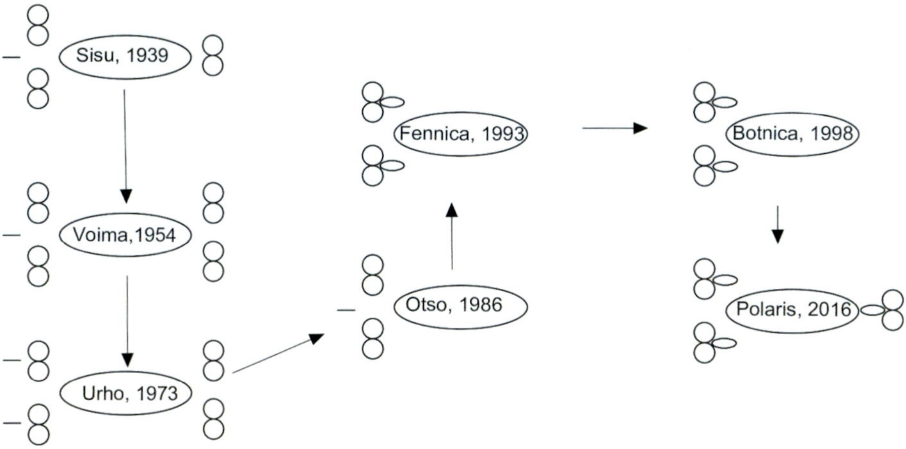

FIGURE 5.18 The history of the propulsion systems for Finnish icebreakers built in Finland.

FIGURE 5.19 The mechanical azimuthing propulsion system on *IB Nordica*.

(Courtesy of Tero Hänninen/Arctic.)

of blade profile on cavitation. The effect of the direction of rotation of bow propellers is also discussed.

The paper presents model tests to determine the wake and thrust deduction coefficients for three icebreaker bow and stern propellers. The wake fractions (potential wake, or the deceleration of the flow aft of the vessel due to the potential wake flow) are quite high, but it should be noted that in two cases the wake fraction for the bow propellers is greater than that of the stern propellers, while in one case it is the reverse. Nonetheless, in all cases the thrust deduction coefficients are higher for the bow propellers.

(a) (b)

FIGURE 5.20 The Azipod azimuthing propulsion system on *IB Polaris*.

(Courtesy of Davie Helsinki Shipyard.)

In addition, a reduction in the ice-hull friction coefficient from 0.15–0.25 to 0.01 is suggested as a result of the flushing effect of the bow thrusters, which is interpreted to be due to ice wetting. According to the study, the primary significance of bow thrusters is the flushing effect, as they are poor thrust generators due to high thrust deduction. The flushing effect is significantly higher with inward rotating bow thrusters because the pressure against the hull is higher, so the optimal combination is inward rotating bow thrusters and outward rotating stern thrusters. Flushing is also more efficient at higher propeller speeds, with smaller bow thrusters.

5.3.2.2 Propulsion Efficiency in Ice (Juurmaa and Segercrantz, 1981)

In addition to the conventional efficiency metrics, a so-called ice efficiency is presented in this study. The ice efficiency is defined as follows:

$$\eta_i = \frac{\eta_{Dice}}{\eta_D} = \frac{(1 - t_{ice})}{(1 - w_{ice})} \cdot \frac{(1 - w)}{(1 - t)} \cdot \frac{T_{ice}}{T} \cdot \frac{V_{Aice}}{V_A} \cdot \frac{Q}{Q_{ice}} \cdot \frac{n_{ice}}{n} \tag{5.25}$$

where the subscript ice refers to tests conducted in ice. Since the speed of the ship is:

$$V = \frac{V_{Aice}}{1 - w_{ice}} = \frac{V_A}{1 - w} \tag{5.26}$$

The expression for the ice efficiency is given by:

$$\eta_i = \frac{(1 - t_{ice})T_{ice}}{(1 - t)T} \cdot \frac{Q}{Q_{ice}} \cdot \frac{n_{ice}}{n} \tag{5.27}$$

The coefficient $(1 - t)T$ can also be considered as net thrust T_{net}, after thrust deduction for a given velocity, rotational speed n and torque Q in open water. Similarly, the coefficient $(1 - t_{ice})T_{ice}$ is the net thrust in ice T_{netice}, for the same velocity, rotational speed and propeller torque Q_{ice}. The ice efficiency is ultimately given by:

$$\eta_i = \frac{T_{netice}}{T_{net}} \cdot \frac{Q}{Q_{ice}} = \frac{1 - \frac{\Delta T}{T}}{1 + \frac{\Delta Q}{Q}} \tag{5.28}$$

where ΔT and ΔQ are the differences in net thrust and torque between open water and ice at the same rotational speed n and velocity v.

The above ice efficiency is relatively straightforward to determine in model tests, but resistance cannot be measured at full scale. Juurmaa and Segercrantz (1981) present three different methods for the approximate determination of the resistance. The first is based on torque from the propeller shaft and propulsion efficiency:

$$R_{Tice} = \eta_i \eta_D \frac{2\pi n_{ice} Q_{ice}}{V} \tag{5.29}$$

The problem is the ice efficiency η_i, which needs to be determined by model tests. The second method is based on measuring the thrust from the propeller shaft:

$$R_{Tice} = (1 - t_{ice}) T_{ice} \tag{5.30}$$

The current problem is the thrust deduction coefficient in ice, which needs to be determined through model tests. The third method is more approximate because it is based on the assumption that the thrust achieved at a given rotational speed in ice is equivalent to the thrust achieved at the same rotational speed in open water. The thrust deduction coefficient is also assumed to be the same in ice and open water. This method can only be used when the propeller-ice interaction is relatively small. This method can be effective even for larger propeller-ice interactions, as the research indicates that the effect of ice on thrust is much smaller than the effect of ice on torque. The advantage of this method is that it can be applied without the need for model test results.

The study presented the results of open-water propeller tests and propeller tests in the ice channel for three different propellers, the ice efficiency for the same propellers in the ice channel were presented.

5.3.2.3 Propulsion Coefficients and Propeller Design for Ice-Going Vessels (Lewis, 1983)

At the beginning of the study, Lewis (1983) suggests that three different categories of information are needed when designing an ice-going vessel:

- Resistance of the bare hull in ice and open water;
- Propeller thrust and torque data in terms of rpm and rotational speed (also pitch if reversible-pith propeller); and
- Information of the interaction between the propeller and the ship's hull, including in ice conditions.

The operation of the propeller of an icebreaker can be compared to the operation of the propeller of a tugboat. According to Lewis (1983), the thrust deduction coefficient, the wake fraction and the relative rotational efficiency are of a completely different magnitude in overload conditions compared to the self-propulsion point in open water. He first discusses the overload condition in open water and then moves on to examine the same parameters in ice.

Lewis (1983) assumes that the thrust deduction coefficient in ice can be determined by:

$$t_i = \frac{T_{Bi} - R_{iw}}{T_{Bi}} \tag{5.31}$$

where T_{Bi} is the thrust in ice, R_{iw} is the total resistance in ice and t_i is the thrust deduction coefficient in ice.

Now, assuming that the thrust deduction coefficient in ice t_i is equal to the thrust deduction coefficient in the overload test in open water t_P, then:

$$R_{iw} = P + R_{ow} \tag{5.32}$$

where P is the towing force caused by ice and R_{ow} is the open-water resistance. The study cannot show that the thrust deduction coefficient in open water in the overload condition is the same as the thrust deduction coefficient in ice. The variability in the results for t_i is so large that no clear conclusion can be drawn.

Lewis (1983) also considers the possibility of determining the total hull resistance in ice in full-scale tests using the propeller efficiency. If the propulsion efficiency in ice is the same as in open water, then the hull resistance is obtained as follows. The idea is to plot the propulsion efficiency obtained from model tests in overload conditions in open water as a function of propeller power for specific propeller rotational speed and ship speed. From the measured propeller power in a full-scale test, the propulsion efficiency is obtained using this curve at the same propeller rotational speed and ship speed. Based on the propulsion efficiency expression and the assumption, $\eta_{Di} = \eta_{DP}$, R_{iw} can be calculated as follows:

$$\eta_{Di} = \eta_{DP} = \frac{P_E}{P_D} = \frac{R_{iw}V}{2\pi nQ} \Rightarrow R_{iw} = \frac{2\pi nQ\eta_{DP}}{V} \tag{5.33}$$

Lewis (1983) investigated the validity of the above assumption and found that the relative rotational efficiency is approximately the same under overload conditions in open water and ice, but the wake fractions are significantly different. The conclusion was that η_{Di} is not as large as η_{DP}, and thus the moment identity for determining ice resistance in full-scale ice resistance tests should not be used. Figure 5.21 shows the K_T and K_Q curves, the relative rotational efficiency and the wake fraction for the ice conditions and the overload condition.

Lewis (1983) also presents a comparison of the differences between the K_T and K_Q coefficients determined for a ship in model and full-scale tests. Figure 5.22 shows that the moment coefficient determined for the propeller behind the hull in the model scale is significantly higher than that determined in the full scale. Conversely, the thrust coefficients obtained from model tests are lower than those from full-scale experiments, but the difference is much smaller.

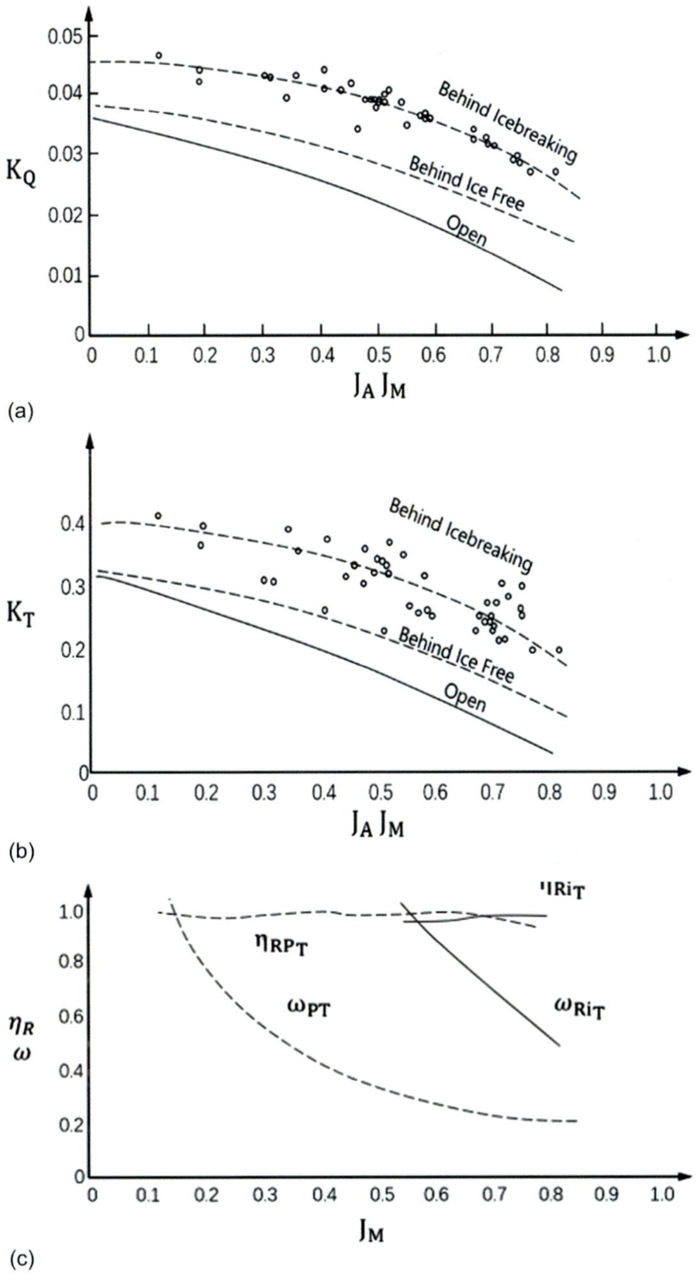

FIGURE 5.21 Thrust coefficient, torque coefficient, relative rotational efficiency and wake fraction in ice and overload conditions (dashed line) and propeller open-water curves.

(Reproduced from Lewis, 1983.)

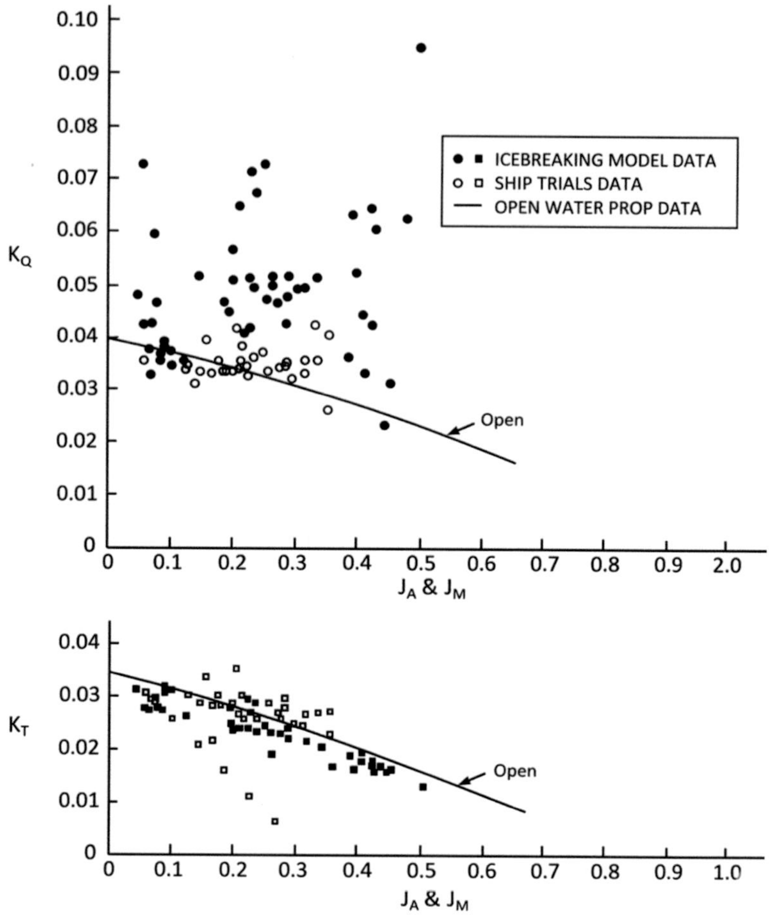

FIGURE 5.22 Thrust and torque coefficient curves from model and full-scale tests and propeller open-water curves.

(Reproduced from Lewis, 1983.)

Lewis (1983) also discusses the design criteria for the propeller of ice-going vessels. In this context, different values of the wake fractions as a function of the advance coefficient are presented. Figure 5.23 shows a typical curve for the wake fraction in overload conditions. The wake fraction w approaches a value of 1/2 as the advance coefficient J approaches 0, and at high values of the advance coefficient, *www* is close to 0.3. Sometimes in the overload condition the thrust coefficient curves for the propeller and the propeller behind the hull intersect at low values of the advance coefficient. In this case, when the wake fraction is determined using the thrust identity, large negative values of the wake fraction are obtained at low values of the advance coefficient. According to Lewis (1983), this illustrates how sensitive the wake fraction value is due to its definition and the shape of the vessel's stern. The study does not address conditions where the propeller operates in front of the hull, so that there is no hull obstructing the wake to change the condition.

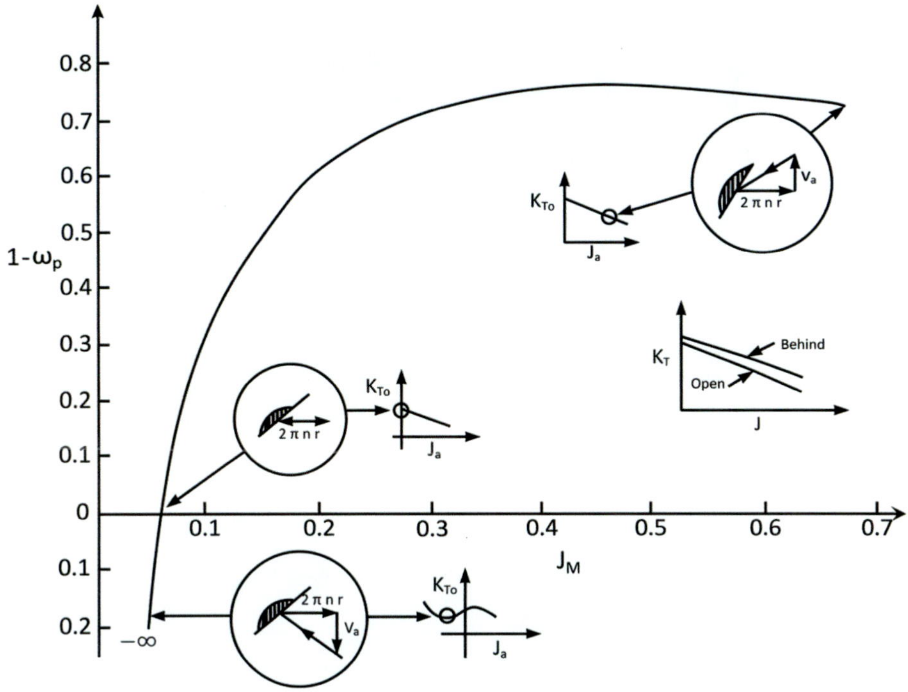

FIGURE 5.23 Typical wake fraction (as a function of 1-w) as a function of the advance coefficient J in towing (overload) conditions.

(Reproduced from Lewis, 1983.)

As the few studies given above show, the presence of ice and the propeller's load near bollard-pull conditions complicate the reliable determination of the propulsion coefficients, resulting in significant scatter in typical experimental measurements, as can also be seen in the case study of extensive model-scale testing for Azipod propulsion.

5.3.3 Pod Propulsion in Ice

In 2001–2002, Helsinki University of Technology carried out an extensive series of tests to study the propulsion coefficients of the *MT Uikku* in forward and reverse in open water and ice on a model scale. The *MT Uikku* was equipped with pod propulsion, see Figure 5.24 (Leiviskä, 2004). The main results of the tests are briefly summarized below.

The purpose of the measurements is to determine, through model test methods, whether the techniques used in open-water model tests can be reliably applied in ice model tests to determine propulsion coefficients and to assess the propulsion coefficient of a vessel equipped with a z-drive in various ice conditions while operating both forward and astern.

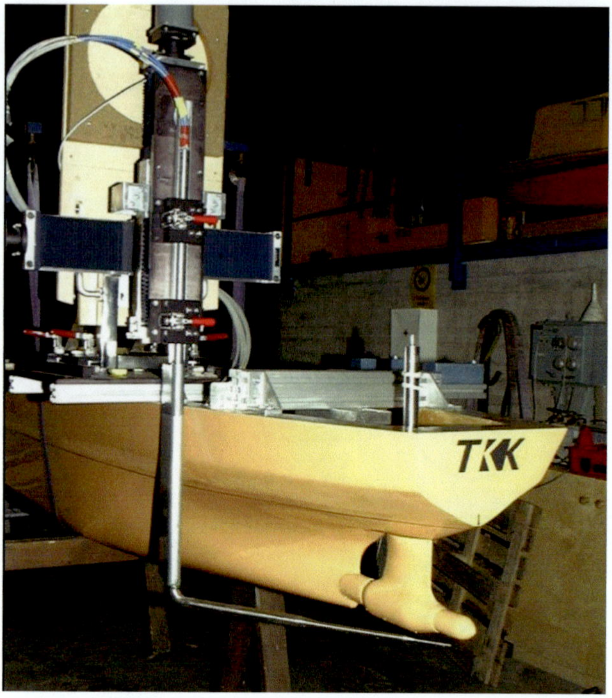

FIGURE 5.24 Model of *MT Uikku* and wake measurement apparatus (Leiviskä, 2004).

The following tests were carried out using the *MT Uikku* model in the model test facilities:

- Propulsor open-water tests forwards and backwards
- Open-water resistance tests of the model forwards and backwards
- Open-water forward and backward overload propulsion tests
- Wake tests backwards
- Propulsion tests on uniform ice at various ice thicknesses, both forward and backwards
- Resistance tests corresponding to the propulsion tests at the same ice thicknesses, both forward and backwards
- Propulsion tests in the ice channel, both forward and backwards
- Resistance tests in the ice channel, both forward and backwards

In addition, tests were conducted at the Arctic Centre of Kværner Masa-Yards (MARC) on a uniform ice field with three different rotational speeds and in the ice channel with one rotational speed, all in the backward direction.

5.3.3.1 Thrust Deduction Coefficient

Figures 5.25 and 5.26 show the measured thrust deduction coefficient in open water, forward and backwards, respectively. Similarly, the measured total thrust deduction

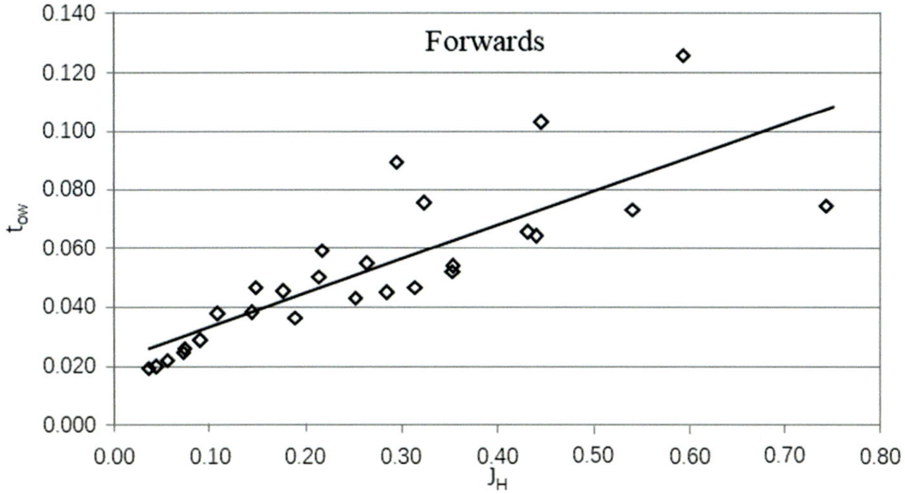

FIGURE 5.25 Thrust deduction coefficient in open water under overload conditions, forwards as a function of the ship's advance coefficient (Leiviskä, 2004).

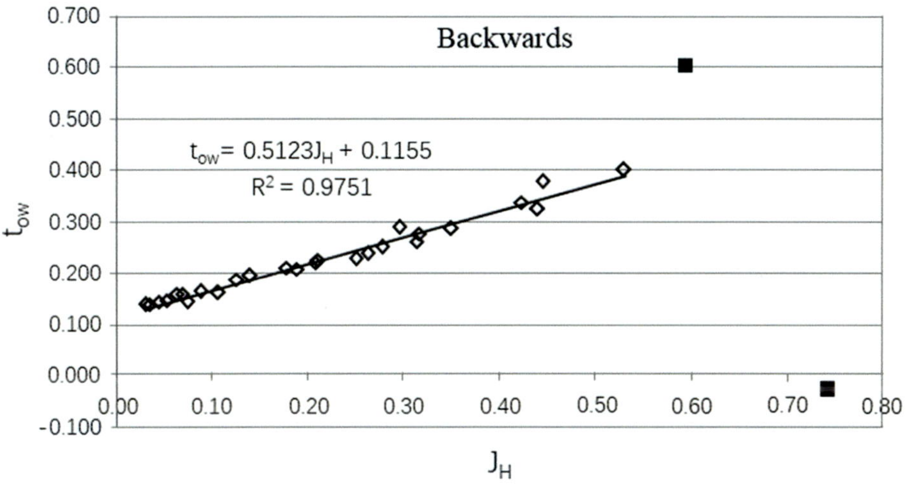

FIGURE 5.26 Thrust deduction coefficient in open water under overload conditions, backwards as a function of the ship's advance coefficient (Leiviskä, 2004).

coefficient in ice are shown in Figures 5.27 and 5.28. Due to the variability in the measurement results of the ice tests, difference margins are also shown for the results. The forward thrust deductions coefficient is of similar magnitude; however, due to variability, the dependence on the advance coefficient in ice is not as straightforward. The backward thrust deduction coefficient in ice is of the same order of magnitude as

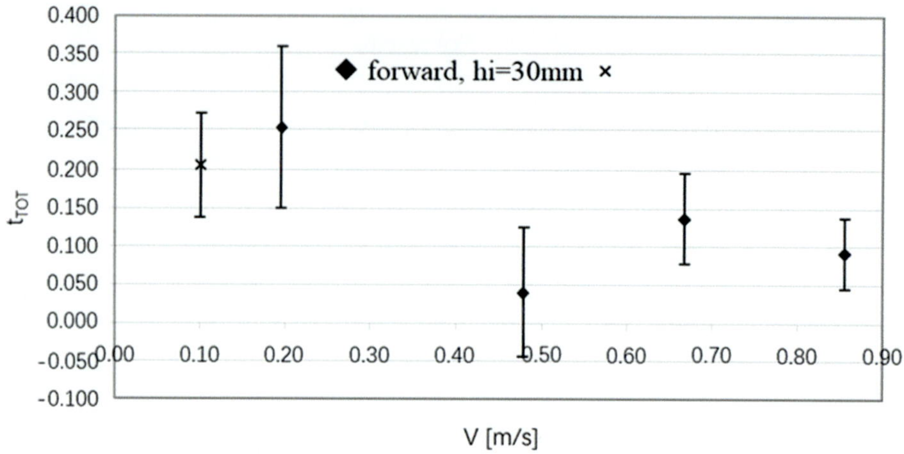

FIGURE 5.27 Total thrust deduction coefficient forwards in level ice and in a ice channel as a function of the ship's advance coefficient, with difference margins (Leiviskä, 2004).

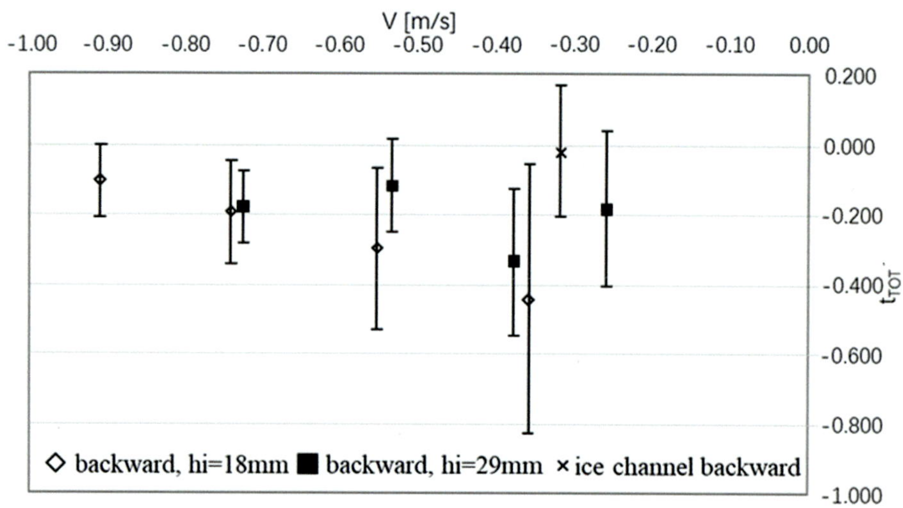

FIGURE 5.28 Total thrust deduction coefficient backwards in level ice and in an ice channel as a function of the ship's advance coefficient, with difference margins (Leiviskä, 2004).

in open water but with the opposite sign. When operating backwards in ice, the thrust deduction coefficient is negative because the propeller flushing effect actually reduces the resistance when operating backwards.

Determining the wake fraction in ice tests has proved difficult. The wake fraction can easily reach negative values and at low advance coefficient; it can exhibit very large positive or negative values.

5.3.4 The Net Thrust Model

There are many parameters affecting ship propulsion performance in polar navigation, among which ice conditions (layer ice and brash ice), brash ice parameters (including brash ice thickness and shape), ship-ice friction resistance coefficient, channel width, brash ice density and ship's speed have an impact on ship propulsion performance. Regardless of how the influencing parameters are changed, ultimately it is the interaction between ice and propeller that leads to the degradation of ship propulsion performance. Therefore, the propulsion performance of polar ships needs to focus on analysing the effect of ice on propellers.

The required propulsion power in the design ice conditions cannot be determined directly from the effective power ($P_E = R_T{}^*v$) or the thrust power ($P_T = T^*v_A$) because the ship speeds are low and the efficiency of the propulsion must be addressed. Thus, the power requirement is derived from the power needed to generate the bollard pull. The derivation of the bollard pull and the power requirement is presented below.

A relationship must be developed between the ice resistance R to be overcome at the design speed and the power to produce this thrust. This can be done by first considering an ideal propulsor. The power delivered at the bollard-pull situation (speed of advance zero) of an ideal propulsor is:

$$P_D = \sqrt{\frac{T^3}{2\rho A_0}} = \sqrt{\frac{2T^3}{\pi\rho D_P^2}} \tag{5.34}$$

where A_0 is the propeller disc area.

From this expression the bollard thrust of an ideal propulsor is obtained as:

$$T = \left(\frac{1}{2}\rho\pi\right)^{\frac{1}{3}} (P_D D_P)^{\frac{2}{3}} \tag{5.35}$$

Similar expression for actual propellers instead of ideal propellers can be obtained using the propeller $K_T - K_Q$-curves as:

$$T = \frac{K_T}{K_Q{}^{2/3}} \sqrt[3]{\frac{\rho}{4\pi^2}}^{-3} (P_D D_P)^{2/3} \tag{5.36}$$

where $\frac{K_T}{K_Q{}^{2/3}}$ is an empirical definition for the quality criterion for bollard pull and is normally denoted as K.

$$K = \frac{K_T}{K_Q{}^{2/3}} \tag{5.37}$$

to be calculated at $J = 0$.

The factor K describes the goodness of the propeller open-water characteristics given the bollard pull. A larger value of K means more thrust for a given power.

The above expressions suggest the definition of a bollard thrust coefficient is the ratio of the actual and ideal thrust as:

$$\eta_{BP} = \frac{K_T}{K_Q^{2/3}} \frac{1}{\sqrt[3]{2\pi}} \approx 0.253 \frac{K_T}{K_Q^{2/3}} = 0.253K \tag{5.38}$$

Usually, however, a dimensional coefficient is used to characterize the bollard pull. This dimensional coefficient is called the quality constant of the bollard pull, K_e, and is defined at the bollard pull situation as:

$$T_{pull} = K_e(P_s D_P)^{2/3} \tag{5.39}$$

where the units of the variables are kN, kW and m, respectively. Note that the K_e-coefficient is dimensional. Therefore, the units of thrust, power and diameter must be as above. The relationship between the dimensionless factors and the quality coefficient is:

$$K_e = K\left(\frac{\rho}{4\pi^2}\right)^{\frac{1}{3}}(1-t) \tag{5.40}$$

where t is the thrust deduction.

The design speed in the Finnish-Swedish ice-class rules is 5 knots, so the bollard-pull value should be complemented with some knowledge about open-water resistance. This is done here by introducing the concept of net thrust T_{net}. This is the thrust available to overcome the ice resistance after the thrust used to overcome the open-water resistance is taken into account. Thus, the net thrust is defined as:

$$T_{net}(v) = T_{tot}(v)(1-t) - R_{ow}(v) \tag{5.41}$$

The thrust deduction is taken into account with thrust deduction fraction t. Even though t is not the same in open-water conditions and ice conditions, but it varies in a complicated manner with speed and resistance; it is simply estimated as a reduction from the total thrust with factor $1 - t$. The effect of speed v is approximated here by a quadratic factor $f(v)$ depending on the maximum open-water speed v_{ow} as:

$$T_{net}(v) = f(v) \cdot T_{pull} = \left(1 - \frac{1}{3}\frac{v}{v_{ow}} - \frac{2}{3}\left(\frac{v}{v_{ow}}\right)^2\right) \cdot T_{pull} \tag{5.42}$$

when the ship speed is zero, the net thrust is equal to bollard pull. When the ship speed is the maximum open-water speed, the net thrust is zero.

The net thrust model is useful when the exact characteristics of the propeller is unknown. If the open-water characteristic curves are known, the net thrust can be calculated alternatively by calculating the difference between propeller thrust using K_T curves (shown in Figure 5.22) and open-water resistance.

5.4 SUMMARY

This chapter gives an extensive overview of ship resistance in level ice, brash ice and ridged ice. The basic, widely used formulations are presented for those ice conditions together with a short physical background for the used formulations. Separate section is devoted to the ship manoeuvrability in ice, which is important especially for icebreakers that should be capable to assist ships on all ice conditions. Ship propulsion in ice and propulsion efficiency is systematically reviewed with emphasis on the revolutionary azimuthing propulsion, which has remarkably improved the ship performance on ice waters.

REFERENCES

Enkvist, E., and Johansson, B. M., 1968. On icebreaker screw design. *European Shipbuilding*, 17(1), 1968.

Enkvist, E., and Mustamaki, E., 1986. Model and full-scale tests with an innovative icebreaker bow. *Society of Naval Architects and Marine Engineers-Transactions*, 94 (Preprint No. 13), pp. 325–352.

ITTC, 2024. Recommended procedures and guidelines 7.5-02-04-02.3: Manoeuvring Tests in Ice. https://ittc.info/media/11838/75-02-04-023 [19.03.2026]

Juurmaa, K., and Segercrantz, H., 1981. On propulsion and its efficiency in ice. In *Proceedings of the Sixth Ship Technology and Research (STAR) Symposium*. The Society of Naval Architects and Marine Engineers, Ottawa, Ontario, 17–19 June 1981, pp. 229–237.

Kashteljan, V. I., Poznyak, I. I., and Ryvlin, A. Y., 1968. *Ice resistance to motion of a ship*. Sudostroyeniye.

Kujala, P., and Riska, K., 2010. Talvimerenkulku, Helsinki University of Technology, Faculty of Engineering and Architecture, Department of Applied Mechanics, Series AM, Report TKK-AM-13. In Finnish.

Leiviskä, T., 2004. *Ship propulsion coefficients in ice and open water*. TKK, Ship Laboratory, Research Report M-287.

Lewis, J. W., 1983. Lecture notes on propulsion coefficients and propeller design for icebreaking ships. In *Western European graduate education in marine technology (WEGEMT) VII graduate school*. Helsinki: Ships and Structures in Ice, pp. 1–38.

Li, F., Lu, L., Puolakka, O., and Kujala, P., 2024. Ice channel breakout performance of a double-acting vessel. *Ocean Engineering*, 293, 116657.

Lindqvist, G., 1989. A straightforward method for calculation of ice resistance of ships. The 10th International Conference on Port and Ocean Engineering under Arctic Conditions, June 12-16, 1989, Luleå, Sweden

Malmberg, S., 1983. *Om fartyg fastkilning I is*. Master's Thesis, HUT.

Matala, R., and Suominen, M., 2023. Scaling principles for model testing in old brash ice channel. *Cold Regions Science and Technology*, 210, 103857.

Mellor, M., 1980. Ship resistance in thick brash ice. *Cold Regions Science and Technology*, 3 (4), 305–321.

Myland, D., 2019. *Experimental and theoretical investigations on the ship resistance in level ice*. Doctoral Thesis. NTNU.

Riska, K., 2017. Ice action on ship hull. In *Encyclopedia of maritime and offshore engineering*. John Wiley & Sons.

Riska, K., Leiviskä, T., Nyman, T., Fransson, L., Lehtonen, J., Eronen, H., and Backman, A., 2001. Ice performance of the Swedish multi-purpose icebreaker Tor Viking II. In *Proceedings of the 16th International Conference on Port and Ocean Engineering Under Arctic Conditions*, pp. 849–865, Ottawa, Canada, 12–17 August.

Riska, K., Wilhelmsson, M., Englund, K., and Leiviskä, T., 1997. Performance of merchant vessels in ice in the Baltic. Winter Navigation Research Board. Report No. 52.

Runeberg, R., 1889. On steamers for winter navigation and ice-breaking (including plates at back of volume). In *Minutes of the Proceedings of the Institution of Civil Engineers*, Vol. 97, pp. 277–301.

Wilkman, G., Hänninen, S., and Heideman, T., 2016. Development of propulsion technology of icebreakers. In *OTC 27462. Arctic Technology Conference, St. John's, Newfoundland and Labrador*, Canada, 24–26 October 2016.

Wilkman, G., Hänninen, S., and Heideman, T., 2019. Development of propulsion technology of arctic ships operating in ice, from MV arctic to Norilskiy Nickel. In *Proceedings of the 25th International Conference on Port and Ocean Engineering Under Arctic Conditions*, Delft, Netherlands, 9–19 June 2019.

6 Polar Ship Rules

6.1 DESIGN PHILOSOPHY

Based on the United Nations Convention on the Law of the Sea (UNCLOS) article 234, coastal states have the right to adopt and enforce non-discriminatory laws and regulations for the prevention, reduction and control of marine pollution from vessels in ice-covered areas within the limits of the exclusive economic zone, where particularly severe climatic conditions and the presence of ice covering such areas for most of the year create obstructions or exceptional hazards to navigation, and pollution of the marine environment could cause major harm to or irreversible disturbance of the ecological balance. Such laws and regulations shall have due regard to navigation and the protection and preservation of the marine environment based on the best available scientific evidence.

Based on UN Convention stated above, some countries have developed ice-class rules to ensure safe shipping on ice-covered waters; the most important ice-class rules have been defined by Finland/Sweden, Canada and Russia. These will be described in further detail below. International Code of Safety for Ships Operating in Polar Waters (Polar Code) came into force in the beginning of 2017. All these regulations will be described later in this section. Figure 6.1 illustrates the geographical areas of the various ice-class rules.

The history of ice-class rule development was started by the first Finnish rules in 1890 having some general guidelines for outfitting and double bottom design. The next Finnish rules (1920, 1932) defined the so-called percentage basis of the scantlings from open water to ice and in 1932 the ice classes, IA, IB and IC, were defined. The first Finnish-Swedish Ice-Class Rules (FSICR) were developed in 1971 in which the ice load was defined for the first time and ice-class IA Super was defined with major updates in 1985 and 1999. The first Russian rules were developed in the early 1960s with major updates in 1995 and 2008. The first Canadian rules (ASPPR) were published in 1972 with major update in 1989. More details on the development will be given for each rule on separate chapters, where these rules are defined in further details, see Section 6.2.

Dimensioning regulations for ships travelling in ice are also issued by most classification societies. Ice classes only become relevant when an authority requires a certain class. Prior to this, the ice classes can only be considered as a dimensioning guide for an educated ship owner to avoid damage.

In Finland and Sweden, and partly in Russia, it is a practice to impose traffic restrictions on sea areas, i.e., there is a minimum ice class and dead weight that is required. In Russia, minimum power is given instead of dead weight. In the Arctic regions of Canada, the limitation comes from the so-called ice numeral per ice class. Canada's ice numerical calculations are based on the risk of environmental damage, i.e., the strength of the ship. In Finland, fairway fees are also linked to the ice class,

DOI: 10.1201/9781003676829-6

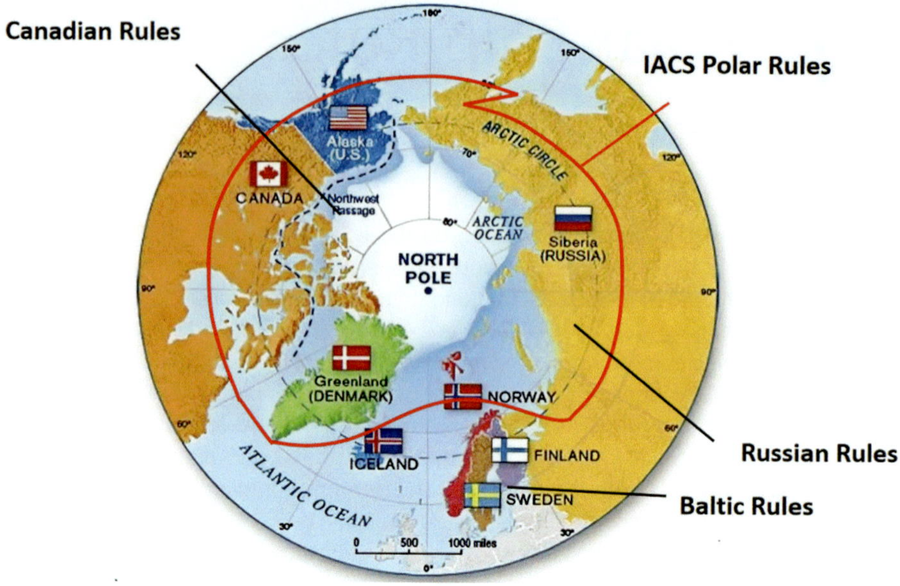

FIGURE 6.1 Ice-class areas.

(Re-produced from Daley, 2014.)

the higher the ice class, the lower the fairway fee. This is why the Finnish regulations are called ice-class payment.

The ice-class rules basically define the required scantlings for the hull and the design requirements for the propulsion system, especially for the propeller blades. Power requirements are only defined in the Finnish-Swedish and Russian ice-class rules. All rules have somewhat different approach to define the loads and the strength criteria. Historically, the damage statistics have had an important role on rule development. Typically, the rule requirements include some physical background and then some empirical coefficients are applied to obtain the right scantlings for the hull and propellers. The full-scale measurements of ice-induced loads on the hull and propulsion systems, as summarized in Chapter 7, have also had important role in developing the rules. It is important to notice that the loads defined in the rules have to be considered together with the used strength criteria as, e.g., FSICR has first yield as a strength criterion and practically all the others rules have the ultimate strength as strength criteria.

Figure 6.2 illustrates typical failure process for the frames of an ice-strengthened ship hull structure. The structural behaviour is elastic until the first yield and thereafter plastic deformations start causing permanent changes in the structure, see also Section 3.3 and Figure 3.24 how the plasticity will increase. The mechanism action in Figure 6.2 denotes a condition where extensive plastic deformation develops in the frame, leading to loss of its load-carrying capacity and attainment of ultimate strength, which is the dominant failure criteria in ice-class rules. The structures still have a lot of

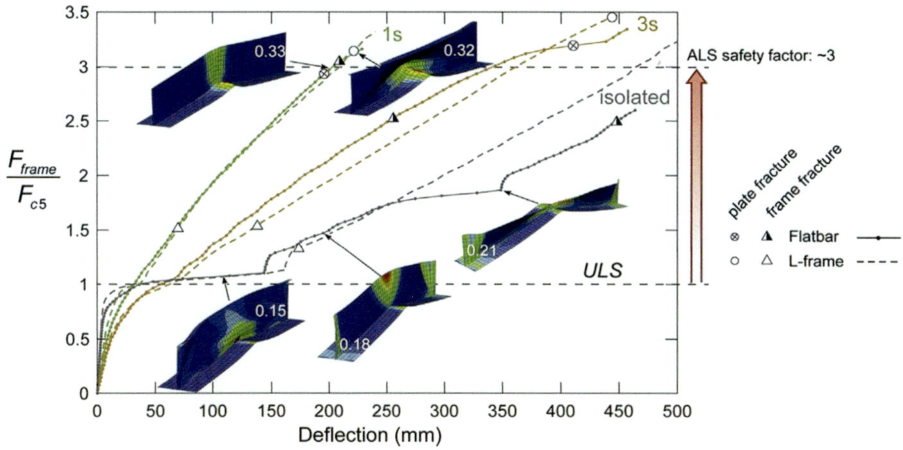

FIGURE 6.2 Normalised frame response to show the safety margin for accidental limit state. Load height is 0.1 m. (a) Flatbars and (b) L-frames. Inset figures show the contours of plastic strains together with the maximum plastic strain in the frame (Körgesaar et al., 2018).

load-carrying capacity left before any possible fracture takes place on the hull causing the water leakage into the ship hull. This behaviour has been systematically studied by Körgesaar et al. (2018) as already described in Section 3.3.

The behaviour of shell plating is somewhat different as after the first yield the plastic deformations will cause large deflections of the plating and so-called membrane forces will increase the load-carrying capacity of the plating (Jones, 1976). Figure 6.3 illustrates typical damage on the shell plating of ice-strengthened ships.

6.2 ICE-STRENGTHENING RULES FOR SHIPS

6.2.1 Finnish-Swedish Ice-Class Rules

6.2.1.1 Definition of Ice Classes

The latest instructions in Finland on how to classify a ship into different ice classes were given in 2005 with minor updates in 2018. This law on ship ice classes and icebreaker assistance issued by the Finnish Parliament clarifies the ships belonging to different ice classes (Law, 2005):

1. for ice-class IA Super, a vessel whose structure, engine power and other characteristics are such that it can navigate in difficult ice conditions, as a rule, without the assistance of an icebreaker;
2. ice-class IA ship whose structure, engine power and other characteristics are such that it can navigate in difficult ice conditions with the assistance of an icebreaker as needed;
3. for ice-class IB, a vessel whose structure, engine power and other characteristics are such that it can navigate in moderate ice conditions with the assistance of an icebreaker as needed;

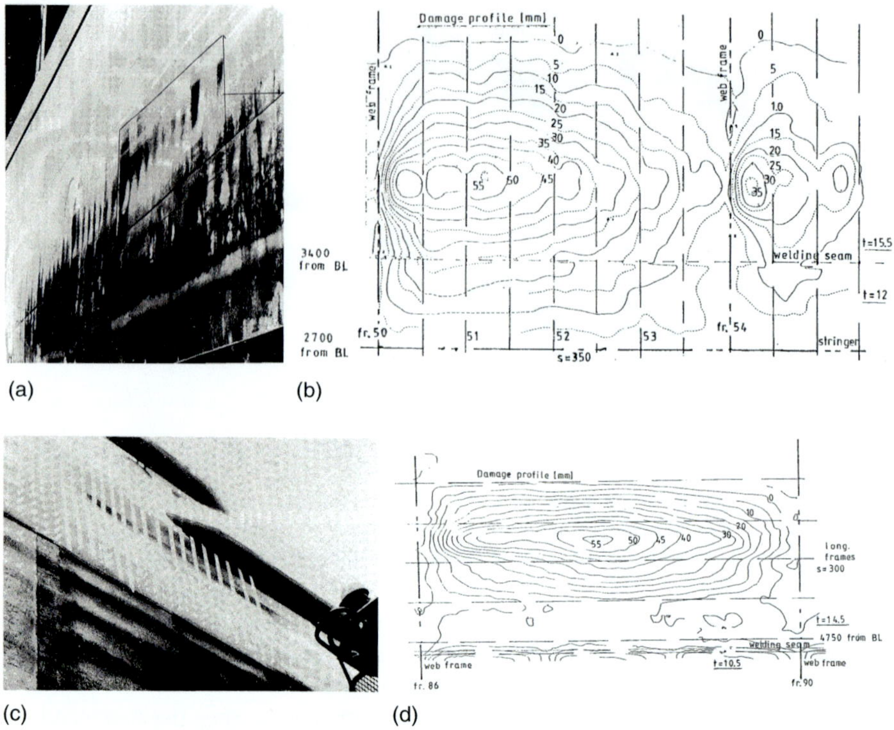

FIGURE 6.3 Examples of typical damages on the shell plating of ice-strengthened ships (Kujala, 1991).

4. ice-class IC ship whose structure, engine power and other characteristics are such that it is able to navigate in easy ice conditions with the assistance of an icebreaker as needed;
5. ice-class II ship, which has a steel hull and is open seaworthy in its construction and which, despite the fact that the platform is not reinforced for navigation in ice, is able to navigate with its own transport machinery in very easy ice conditions;
6. a vessel in ice-class III that does not belong to the ice class referred to in paragraphs 1–5.

The traffic restrictions are given based on the ice condition data received from the icebreakers, based on the authorities' understanding of the traffic capability of each ice class. Ship owners could also make this estimate directly from the ice conditions. In the Baltic Sea, therefore, traffic restrictions are mainly based on the ship's ability to operate in ice.

6.2.1.2 Background for the FSICS

Riska and Kämäräinen (2011) gives a good summary of the development of the FSICS. The first Finnish ice-class regulations were issued in 1890. These rules only contained requirements for ship equipment and the general and structural arrangements on the ship. The following rules were the so-called percentage rules (1920, 1932), i.e., the substance strengths were only increased in percentage from open water. The rules of 1920 required a 45% thickening of the siding board, the installation of intermediate arches and a longitudinal stiffener spacing of 1.8 m (this limited the span of the arches). For the 1932 rules, three different ice classes, IA, IB and IC, were defined as well as ice-class II corresponding to open water ships and ice-class III corresponding to barges. The ice-class IA Super was defined for the first time in 1965.

The 1971 rules (FSICR, 1971) were based on a large amount of data on ice damage. In the 1960s, Johansson (1967) collected ice damage statistics and analysed with the simple plastic design approach using three plastic hinges on how much force was needed to cause the damage that appeared in the statistics, and he also calculated the force for the non-damaged ships included in the statistics. The initial assumption was the height of the ice load, which was set equal to the ice thickness, i.e., 800 cm, as well as the maximum possible ice pressure with different structural elements. Johansson presented the calculated load for each ice damage (in different ice-class categories) as a function of the ship's displacement and engine power as shown in Figure 6.4. By drawing a line at the level below which the damaged ships remain and above which the undamaged ships remain, he concluded the appropriate load level for ice ships in the Baltic Sea. This formed the basis for the 1971 ice rules. These were the first Finnish-Swedish rules and are considered the first modern ice rules, defining directly the ice-induced loads.

The 1971 rules had three parts: ship's power requirement, hull strength requirements and propeller and propeller shaft strength requirements. Ice pressure was a

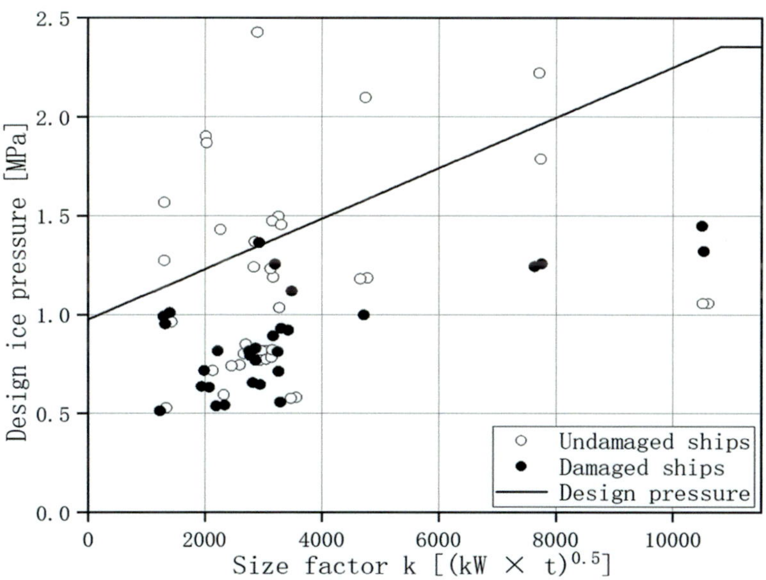

FIGURE 6.4 Ice pressure causing the observed ice damage and the strength of undamaged ships for ice-class IA ships at the bow.

(Re-produced from Johansson, 1967, Riska and Kämäräinen, 2011.)

function of ice class and dependent on the ship's engine power and draft. In the 1971 rules, there were four ice classes: IA Super, IA, IB and IC.

The power requirement is given so that the ships would be able to follow the icebreakers and not slow down the traffic. In the 1971 rules, the strength of the propeller is based on the load acting on the tip of the platform and beam theory. The strength of the shaft is based on the ice moment of the form m•D^2, where m is the coefficient specific to the ice category and D is the diameter of the propeller. Bending resistance is also calculated for the propeller blade. Its value depends on the breaking strength of the blade material, shaft power, number of blades, propeller speed and propeller pitch. Once the propeller blades have been dimensioned, the shaft axis can be dimensioned. According to the dimensioning rules, the shaft axis is strongest the closer it is to the main engines, and the weakest point of the shaft axis is close to the propeller blade.

6.2.1.3 Current Finnish-Swedish Ice-Class Rules for the Hull

The full-scale measurements onboard *IB Sisu* and some other ships in the early 1980s has shown that the load height is remarkably less than the ice thickness as assumed before, see Figure 6.5 (Riska and Kämäräinen, 2011). The high number of damages on the longitudinally framed ships indicated the same (Kujala, 1991). Consequently, for the ice-class rules of 1985 (FSICR, 1985), the load level was kept about the same as in 1971, but the ice-induced pressures were remarkably increased and the load height decreased. This has a remarkable effect especially on the plate thickness requirement for longitudinally framed ships. In addition, the formulations for the plate thickness were updated using the same approach as it is used to Ro-Ro ship decks under wheel loading, assuming some permanent deformation is allowed on the shell plating.

The strength level in each ice class corresponds roughly to the loading from a certain level of ice thickness: 1.0 m for IA Super down to 0.4 m for IC. The ice thickness for IA Super is higher than the maximum level of ice thickness observed in the Baltic outside the fast ice zone. The design point for each ice class is a collision with a channel edge (as ships are escorted in the worst ice conditions) or with a consolidated layer of older ridges. As the consolidated layer of ridges may be 1.8 times thicker than the level ice at the same location, and the maximum average level ice thickness in the middle of the sea basins is about 60 cm, this results in about 1.0-m thick ice (Riska and Kämäräinen, 2011)

The definition of the ice load is the most important part of the hull rules. The ice load in the FSICR is, in principle, defined so that the ice pressure is constant for

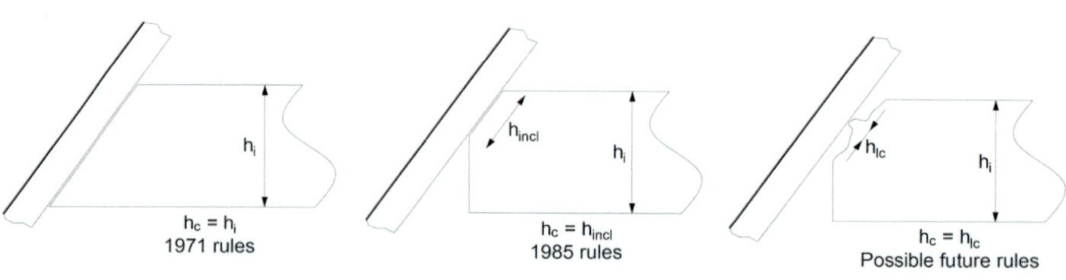

FIGURE 6.5 Sketch of the concept of load height on a ship frame with different rule sets. In all these cases, the value of the line load q has been constant, at about $q = 2$ MN/m.

(Re-produced from Riska and Kämäräinen, 2011.)

all classes (nominal ice pressure p_0) and the load height h is the class factor (from 0.35 m for ice-class IA Super to 0.22 m for ice-class IC). The magnitude of the nominal ice pressure is 5.6 MPa. This approach of constant pressure is taken as the material properties of the Baltic ice do not change much through the winter in different Baltic Sea areas. The total ice load for each structural member is then taken as the line load, q:

$$q = ph \tag{6.1}$$

times a load length l_a that depends on the width (horizontal span or spacing) of each structural member. For transverse frames the load is, for example, $F = q \cdot s$, where s is the frame spacing.

To obtain the ice pressure p, the nominal ice pressure is modified by three multiplicative coefficients, all of which are less than one. The ice damage study carried out in the late 1960s, which resulted in the 1971 rules, indicated that the ship size influences the ice load, see Figure 6.4. An analysis of the mechanics of collision between ice and a ship suggests that it is the ship displacement and speed that influence the contact force. The total contact force in collisions with ice is not an important factor for most of the structural members (they are sensitive to a load patch that is smaller than the total load patch size). This has led to the definition of the size quantity of:

$$k = \sqrt{P\Delta} \tag{6.2}$$

and a size coefficient for ice pressure cd, which is linearly dependent on k. The value of k for a typical Baltic ship (P = 4 MW, Δ =15000 t) is k = 7.7 and the size coefficient cd = 0.46 (bow region) (Riska and Kämäräinen, 2011).

The ship hull is divided into the three hull regions shown in Figure 6.6: bow, midbody and stern. Each of these has a design ice pressure defined by a hull region factor c_p. This factor is for the bow region and is scaled according to the ice class for other regions so that the stern region has the lowest design ice pressure.

The third factor used to define the ice pressure is a coefficient dependent on the load length, c_a. Each structural member has an associated load length l_a – this is the

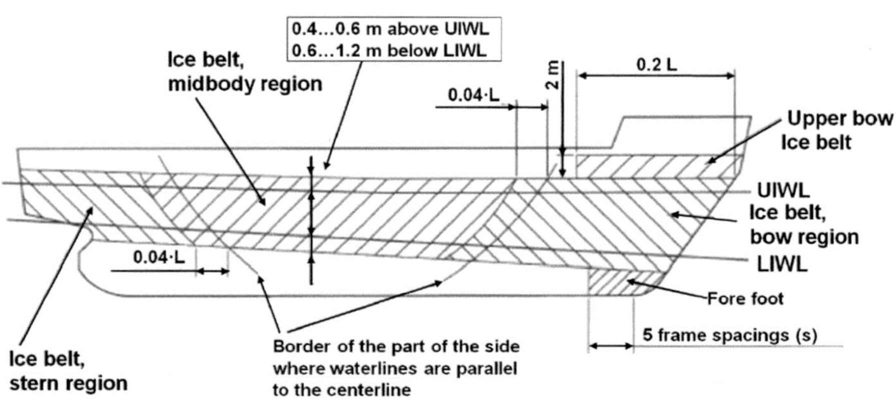

FIGURE 6.6 The hull regions defined in the FSICR.

length of the load that influences the response (stress) in the member. The load length coefficient is defined as:

$$c_a = \sqrt{\frac{l_0}{l_a}} \text{max } 1.0 \text{ and min } 0.35 \tag{6.3}$$

where the reference length is $l_0 = 0.6$ m. In principle, each structural member should be designed by trying all load lengths and then selecting the design case to be the length that gives the maximum stress. There is no need to do this calculation, as the load lengths are given in the rules, as shown in Table 6.1.

The design ice pressure is finally:

$$p = c_d c_p c_a p_0 \tag{6.4}$$

Even if the design ice pressure at the bow region is the same for a ship in all ice classes, the design force is not, as the load height is a class factor.

Once the ice load is specified and the limit state is defined, the scantlings can be calculated. The limit state used for frames in the FSICR is the yield limit – consequently, only the elastic response of the structures needs to be derived. The plate thickness equations are based on a similar equation used to design car decks under tire loading allowing some permanent deformations. The frame equations are based on simple beam formulation. The effect of load height on plate response is taken into account with a constant dependent on h/s (the constant is denoted here as $f(h/s)$) – plate thickness t is given by equations in the form:

$$t = \frac{2}{3} s \sqrt{f\left(\frac{h}{s}\right) \frac{p_{pl}}{\sigma_y}} \tag{6.5}$$

where p_{pl} is the equivalent plate pressure, at 0.75 p, and σ_y is the yield strength. The origin of the constant 0.75 is in the pressure distribution across the plate and is discussed in Section 3.3. The constant $f(h/s)$ is different for transversely and longitudinally framed structures.

The frame scantlings include the frame section modulus Z and shear area A using first yield as the strength criteria for the frames. These are calculated for transverse frames with equations of the form:

TABLE 6.1
Load Lengths Associated with Different Structural Members

Structural Member	Type of Framing	Design Load Length l_a (m)
Shell plating	Transverse	Frame spacing
	Longitudinal	1.7 · frame spacing
Frames	Transverse	Frame spacing
	Longitudinal	Span of frame
Ice stringer		Span of stringer
Web frame		2 · web frame spacing

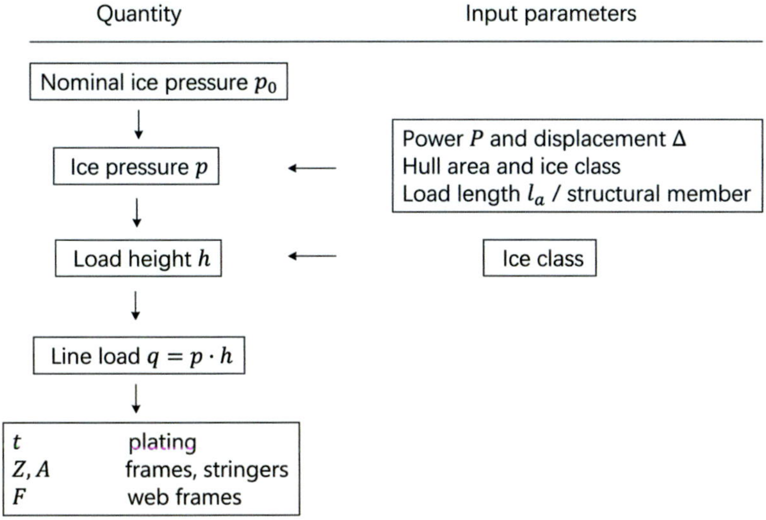

FIGURE 6.7 The structure of the hull rules in FSICR (Riska and Kämäräinen, 2011).

$$Z = \frac{qsl_a}{m\sigma_y} \tag{6.6}$$

$$A = \frac{1}{2}\frac{1.2ql_a}{\tau_y} \tag{6.7}$$

where τ_y is the shear strength and the factor 1.2 stems from taking the shear stress distribution across the web into account. m is a factor dependent on the end connections of the frame ($m = 5 \ldots 7$). The equations for longitudinal frames are similar but contain a factor that is dependent on the load height and frame spacing.

The structure of the hull rules is presented schematically in Figure 6.7.

6.2.1.4 Current Finnish-Swedish Ice-Class Rules for Power Requirements

The starting point of the latest Finnish-Swedish ice rules is the set of ship resistance formulations for navigating in a channel behind an icebreakers, and the ship's rut resistance at a speed of 5 knots in the ice channel described by Equation 5.22 (Riska et al., 1997). Assuming that at such a low speed, the ship's entire resistance is practically due to ice and assuming that T_{net} is approximately 80% of the TB value at this speed, formula (6.8) is obtained in the form:

$$P_D = \frac{R_{ice}^{\frac{3}{2}}}{D}\,(\mathrm{kW}) \tag{6.8}$$

where K_p now gets the following values: for a single-propeller ship with rotating blades, $K_p = 2.03$; for a twin-propeller ship, $K_p = 1.44$; and for a three-propeller ship, $K_p = 1.18$. In Eq. (6.8), R_{ice} is the channel resistance and D is propeller diameter.

For a fixed-blade propeller, these values must be divided by a factor of 0.9. Constant K_p is largely based on experimental measurements and thus includes the effects of thrust reduction and wake coefficients. In practice, these are affected by the shape of the ship's hull, the diameter of the propeller and the required thrust (Juva and Riska, 2003), see also Section 6.3. The effects are so complex that they have not been included in the requirements of the ice rules.

Merchant ships are required to have a minimum performance in ice, so that the icebreaker assistance does not suffer too much due to ships that are slow and constantly needing to be cut out. The merchant ship must be able to follow the icebreaker in a broken channel and move independently even in heavy shoals.

The performance requirement is set in the Finnish-Swedish ice payment class regulations so that, roughly, IA Super ships can move to all ports in the Baltic Sea, IA ships to all ports in the Gulf of Finland and, in an average winter, also in the Bay of Bothnia. IB and IC ships are only easy to navigate in ice and these categories are not recommended. The requirement for the ability to navigate through ice is based on the ship's minimum speed of 5 kn on the following thickness of channel ice.

- IA Super, 1 m of channel ice and 10 cm of frozen level ice
- IA 1 m of channel ice
- IB 0.8 m of channel ice
- IC 0.6 m of channel ice

The ultimate goal, combined with performance requirements and traffic restrictions, is the most efficient transport performance, actually optimizing the transport utility ratio. However, this task is quite difficult, and that is why we ended up with the system currently in use.

6.2.2 Russian Ice-Class Rules

6.2.2.1 Definition of Ice Classes

As stated above, the first Russian rules were developed in the early 1960s with major updates in 1995 and 2008. As also discussed in Section 6.4, Russian scientists have been very active to define the ship-ice interaction formulations, see e.g. Popov et al. (1967). For the 2008 ice-class rules (RMRS, 2010), they have a total of 13 ice classes: three for non-Arctic ice conditions, six for Arctic operations and four for icebreakers. In the following we will only describe the requirements for the Arctic operations.

Table 6.2 shows how the ice classes are defined. The basic idea is to define the ice class based on the navigation sea area along the Northern Sea Route (NSR) either on the winter/spring or summer/autumn time. In addition, the type of navigation is included, i.e., whether the vessel operates independently or under icebreaker assistance and whether the navigation conditions are extreme, hard, medium or easy.

Table 6.3 defines the maximum operative ice thickness for various ice-classes ARC4 to ARC 9. ARC4 is the lowest ice class and ARC9 highest ice class.

TABLE 6.2

The Navigation Areas and Types for Various Russian Ice Classes (RMRS, 2010)

		Winter/Spring Navigation					Summer/Autumn Navigation				
Category of Ice Strengthening	Type of Ice Navigation	The Barents Sea ENHME	The Kara Sea[a] ENHME	The Laptev Sea ENHME	The east Siberian Sea ENHME	The Chuckchee Sea ENHME	The Barents Sea ENHME	The Kara Sea ENHME	The Laptev Sea ENHME	The east Siberian Sea ENHME	The Chuckchee Sea ENHME
Arc4	IN	––––+	––––+	––––+	––––+	––––+	++++	––––+	––––+	––––+	––––+
	IEN	––––+	––––+	––––+	––––+	––––+	++++	++++	––––+	––––+	––––+
Arc5	IN	––––+	––––+	––––+	––––+	––––+	++++	––––+	––––+	––––+	––––+
	IEN	++++	––––+	––––+	––––+	––––+	++++	++++	++++	++++	++++
Arc6	IN	++++	––––+	––––+	––––+	––––+	++++	++++	––––+	––––+	––––+
	IEN	++++	+++	+++	+++	+++	++++	++++	++++	++++	++++
Arc7	IN	++++	––––+	––––+	––––+	––––+	++++	++++	++++	++++	++++
	IEN	++++	++++	+++	+++	+++	++++	++++	++++	++++	++++
Arc8	IN	++++	++++	––––+	––––+	+++	++++	++++	++++	++++	++++
	IEN	++++	+++	+++	+++	+++	++++	++++	++++	++++	++++
Arc9	IN	++++	++++	++++	++++	++++	++++	++++	++++	++++	++++
	IEN	++++	+++	++++	++++	++++	++++	++++	++++	++++	++++

Note: Symbols: IN – Independent navigation, IEN –Icebreaker escorted navigation, + – operation allowed, – – operation not allowed, * – operation connected with the increased risk of damage, EN – extreme navigation (average periodicity once in ten years), H, M, E – hard, medium, easy navigation (average periodicity once in three years).

[a] For ships with ice category mark Arc7 an independent (IN) year-round navigation in the southwestern part of Kara Sea, at EN, H, M and E typed of navigation.

TABLE 6.3
Allowable Ice Thickness for the Summer and Winter Period for Various Ice Classes (RMRS, 2010)

Ship Category	Permitted Type and Thickness of Ice	
	Winter/Spring Navigation	Summer/Autumn Navigation
Arc4	Thin first-year	Medium first-year up to 0.9 m
Arc5	Medium first-year up to 0.8 m thick	Medium first-year
Arc6	Medium first-year	Thick first-year up to 1.5 m
Arc7	Thick first-year up to 1.8 m	Second-year
Arc8	Multiyear up to 3.4 m	Multiyear
Arc9	Multiyear	Multiyear

Note: The classification of ice adopted according to the 'Sea Ice Nomenclature' of the World Meteorological Organization (WMO).

Ice type	Ice thickness
Multiyear	>3.0 m
Second-year	>2.0 m
Thick first-year	>1.2 m
Medium first-year	0.7–1.2 m
Thin first-year	<0.7 m

6.2.2.2 Current Russian Ice-Class Rules for the Hull

There have been a number of minor updates on the Russian rules and here we summarize the current version (RMRS, 2014). Figure 6.8 illustrates how the ship hull parameters at the bow area are defined by RMRS.

Table 6.4 defines the allowable values for the hull shape parameters shown in Figure 6.8. These angles are mainly defined for the bow region, except for ice ARC8 and ARC9, where a minimum frame slope angle of 15° is specified.

In the afterbody of icebreakers and Arc4–Arc9 ice-class ships, there shall be an appendage (ice knife) aft of the rudder to protect the latter on the sternway. No transom stern (with the transom coming in the region of ice strengthening) is permitted for icebreakers and Arc6–Arc9 ice-class ships. For Arc4 and Arc5 ice-class ships, the transom aft end is not recommended in the regions of ice strengthening. In ships of ice-classes Arc8 and Arc9, bulbous bow is not permitted. Figure 6.9 and Table 6.5 illustrate the ice-strengthening areas for RMRS.

A grillage may include vertical girders of main framing which are denoted as conventional frames, vertical web members which are denoted as deep frames, and longitudinals are denoted as stringers.

Conventional frames are subdivided into: main frames in plane of floors or bilge brackets; intermediate frames not in plane as floors or bilge brackets. The intermediate

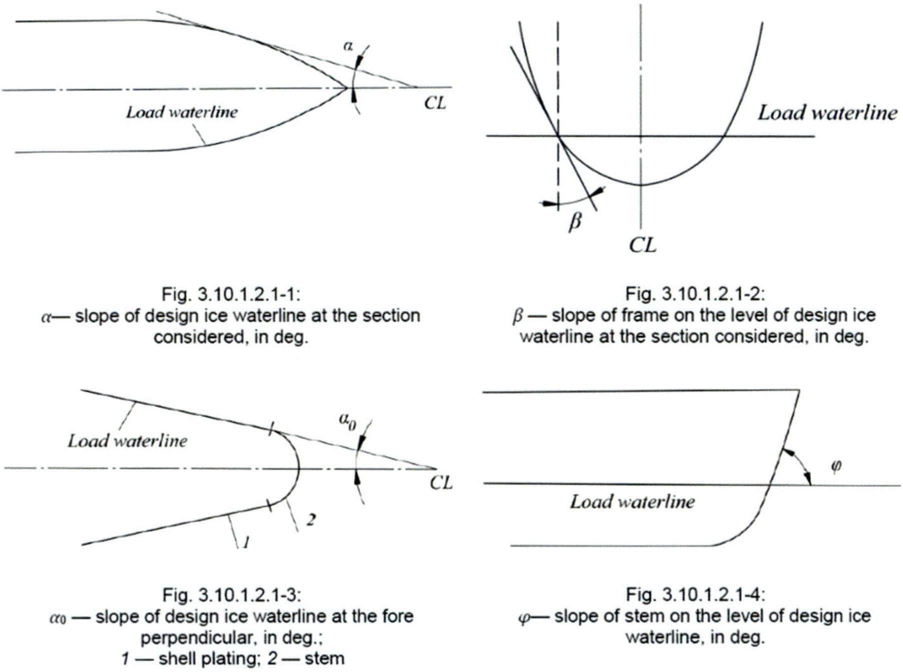

FIGURE 6.8 The definition of the main hull shape parameters (RMRS, 2014).

TABLE 6.4
Allowable Hull Shape Angles for Various Ice Classes (RMRS, 2014)

	Ice Class			
Hull Configuration Parameter	**Arc8, Arc9**	**Arc7, Arc6**	**Arc5**	**Arc4**
ϕ not greater than	25°	30°	45°	60°
α 0 not greater than	30°	30°	40°	40°
β within 0.05 L from fore perpendicular, minimum	45°	40°	25°	20°
β amidships, minimum	15°	——	——	——

frames are not mandatory within a side grillage. Not more than one intermediate frame may be fitted between main frames.

Stringers are subdivided into: intercostal stringers by which joint taking-up of local ice loads by the frames is ensured. It is recommended that the stringers shall be intercostal; side stringers by which a transition of forces is ensured from conventional frames which directly take up the ice load to deep frames or to transverse bulkheads.

Side grillage structures are permitted as follows: grillage with transverse main frames, which is formed by conventional frames of the same section and by intercostal stringers; grillage with transverse web-frames which is formed by conventional frames, side stringers and deep frames. Intercostal stringers may be fitted together with side stringers.

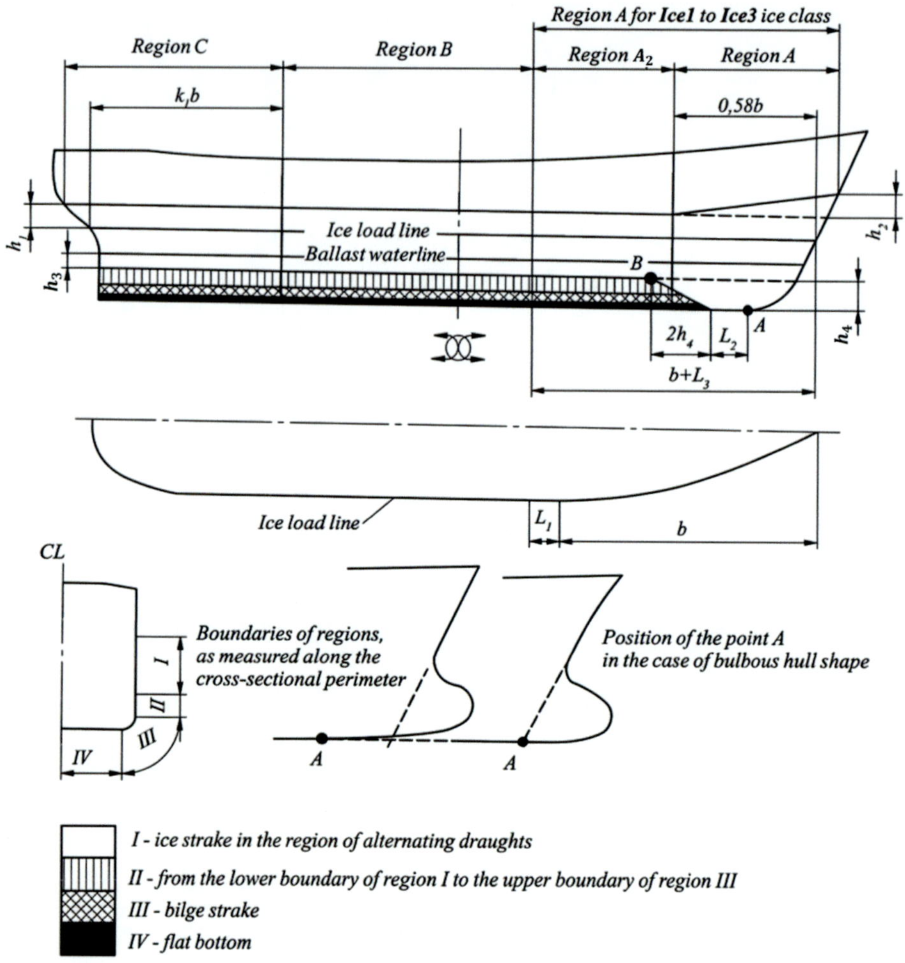

FIGURE 6.9 The hull regions defined in the RMRS. b – distance from the point of the ice loadline and stem intersection to the section where the ice loadline is the widest, but not greater than 0.4 L.

The rules also give detail instructions on how to plan the joining elements of the frames and other structural elements as well maximum spacing of the elements.

The ice load depends on three parameters:

p – ice pressure being a characteristic of the maximum pressure in the area of dynamic contact between the hull and ice, in kPa;

b – vertical distribution of ice pressure characterizing the maximum transverse dimension of the dynamic contact area between the hull and ice, in m;

lp – horizontal distribution of ice pressure characterizing the maximum longitudinal dimension of the dynamic contact area between the hull and ice, in m.

TABLE 6.5
The Hull Region Requirements Defined in the RMRS

Parameter		Ice Class					
		Arc7, Arc8, Arc9	**Arc5, Arc6**	**Arc4**	**Ice3**	**Ice2**	**Ice1**
h_1, in m	Where B B<20 m	0.75		0.60	0.50		
	Where B > 20 m	$\frac{0.5B + 8}{24}$		$\frac{0.5B + 8}{30}$	$\frac{0.5B + 8}{36}$		0.50
h_2, in m		1.4	0.8	0.6	0.2		
h_3, in m		1.6 h_1	1.35 h_1	1.20 h_1	1.10 h_1	h_1	
L_2, in m		0.15 L	0.1 L	0.05 L	0.02 L		—
L_3, in m		0.06 L	0.05 L	0.045 L	0.04 L	0.02 L	
k_1		0.84	0.69	0.55	0.53	0.50	—

The ice pressure is defined separately for various longitudinal areas A, B and C and then again for vertical areas I, II, III and IV, as shown in Figure 6.9. For ice-load parameters for the bow areas with a vertical side ($\beta < 5$) in the region of alternating draughts, or bulbous bows, separate formulations are given, but in the following we will only present the equations for the hull fulfilling the hull form requirements in Table 6.3.

For the area A_I, the pressure is determined as :

$$p_{AI} = 2500 a_1 v_m \sqrt[6]{\frac{\Delta}{1000}} \tag{6.9}$$

where

a_1 is a factor based on the ice class (e.g. 0.79 for ARC4 and 7.9 for ARC9)
Δ is a displacement to the ice loadline, in t;
v_m is a value of the shape factor v, which is the maximum for a region, as determined at sections within $x = 0$; $0{,}025L$; $0{,}05L$; $0{,}075L$; $0{,}1L$, etc. from the forward perpendicular of the design ice waterline. The value shall be determined by the following formulae for ships and icebreakers with the hull shape compliant in Table 6.3:

$$v = \left(0{,}278 + 0{,}18\frac{x}{L}\right)\sqrt[4]{\frac{\alpha^2}{\beta}} \text{at} \frac{x}{L} \leq 0.25 \tag{6.10}$$

$$v = \left(0{,}343 - 0{,}08\frac{x}{L}\right)\sqrt[4]{\frac{\alpha^2}{\beta}} \text{ at} \frac{x}{L} > 0.25 \tag{6.11}$$

where

- L is a length at design ice waterline;
- x is a distance between the considered section and the forward perpendicular, in m;
- α is an angle of design waterline inclination as shown in Figure 6.8 in deg.;
- β is an angle of frame inclination at ice loadline on the considered section as shown in Figure 6.8 in deg.

For regions II, III and IV, the ice pressure is determined as a part of the pressure in region I at the appropriate section of the ship length, by multiplying the value at region I with parameters defined separately for each ice class and region. This value is varying from 0.15 to 0.7, with highest value at the bow with high ice class and lowest at aft with lowest ice class.

The vertical height of ice pressure, in m, shall be determined by the following formula for region A including all the areas I–IV (similar equation also exist for other region A_1, B, C):

$$b_A = C_1 k_\Delta u_m \tag{6.12}$$

where C_1 is a factor defined for each ice class (0.49 for ARC4 and 0.64 for ARC9),

$$k_\Delta = \sqrt[3]{\frac{\Delta}{1000}}$$

u_m is the maximum value of shape factor u for the region, as determined at sections within $x = 0; 0.025L; 0.05L; 0.075L; 0.1L$, etc. from forward perpendicular for ships with the hull shape compliant as shown in Table 6.3:

$$u = k_B\left(0.635 + 0.61\frac{x}{L}\right)\sqrt{\frac{\alpha}{\beta}}\text{at}\frac{x}{L} \leq 0.25 \tag{6.13}$$

$$u = k_B\left(0.862 - 0.30\frac{x}{L}\right)\sqrt{\frac{\alpha}{\beta}}\text{at } \frac{x}{L} > 0.25 \tag{6.14}$$

where $k_B = \begin{cases} 1 \text{ at } \beta \geq 7^o \\ 1.15 - 0.15\frac{\beta}{7} \text{ at } \beta < 7^o \end{cases}$

Horizontal length of ice pressure, in m, shall be determined by the following formulae for ships of ice classes ARC4–ARC7:

$$l_A = 0.337 l^i_{ref} u_{l_m} \left(\frac{\Delta}{1000}\right)^{0.2906} \tag{6.15}$$

where l^i_{ref} is a factor varying for each ice class (4,55 for ARC4 and 4,23 for ARC7), u_{l_m} is maximum value of the shape factor u_l to be determined in sections of bow area with the vertical side at ice loadline by a second-order regression formula with six different parameters given on a table for each ice class.

Once the ice pressure and contact dimensions have been determined, the rule will define the formulation to be used for obtaining the required scantlings for the structural elements. In regions of ice strengthening, the shell plating thickness s_{sp}, in mm, shall not be less than as determined by the formula:

$$s_{sp0} + \Delta s_{sp0} \tag{6.16}$$

where

$$s_{sp0} = 15,8a_0\sqrt{\frac{p}{R_{eH}}},$$

$$\Delta s_{sp0} = 0,75Tu,$$

$$a_0 = \frac{a}{1 + 0,5\frac{a}{c}},$$

where p is an ice pressure in the region under consideration as stated above,

- $c = b$ where the grillage is transversely framed in the region under consideration. In this case, c shall not be greater than the spacing of intercostal stringers or the distance between plate structures;
- $c = l$ where the grillage is longitudinally framed in the region under consideration;
- b is vertical height of ice pressure in the region under consideration as stated above in m;
- l is distance between adjacent transverse members, in m;
- a is spacing of shell plating stiffeners, in m;
- T is planned ship life, in years;
- u is annual reduction of shell plating as a result of corrosion wear and abrasion, in mm per year, varying for ice class and region, being 0.30–0.40 for ARC4–ARC7 at the bow and 0.20–0.28 for midship/aft region. When taking measures to protect the shell plating from corrosion wear and abrasion complying the definitions given in the Part XIII 'Materials' and Part III 'Technical Supervision during Manufacture of Materials' of the Rules for Technical Supervision during Construction of Ships and Manufacture of Materials and Products for Ships, the value u may be reduced by 25% when applying Class II protective coatings and by 50% when applying protective coatings of Class I.

The required geometrical characteristics of frames and girder structure cross-sections, such as the ultimate section modulus W and the web area A, are based on the ultimate strength criterion, and for these RMRS (2014) give detail guidelines on how to determine the required scantlings.

6.2.3 IACS PC Rules

6.2.3.1 Definition of Ice Classes

The International Association for Classification Societies (IACS, 2019) drew up harmonized rules on the basis of the different arctic ice classes that already exist to clarify the ice code, first published in 2007. The goal of IACS is to draw up rules that determine the minimum strength, and the classification institute can, after meeting the IACS rules, demand higher strengths if it feels it is necessary. It is expected that unified PC rules will dominate in future the ice-strengthening principles as the new Polar Code is based on those, see Section 6.4. Transport Canada previously defined its own ice classes, so-called CAC1-4 and ship types A to E, but has since moved to adopting the PC classes on its waters.

For ships which are assigned a Polar Class notation, the hull form and propulsion power are to be such that the ship can operate independently and at continuous speed in a representative ice condition, as defined in Table 6.5 for the corresponding Polar Class. For ships and ship-shaped units which are intentionally not designed to operate independently in ice, such operational intent or limitations are to be explicitly stated in the Certificate of Classification or equivalent.

The IACS (2019) code consists of a total of seven ice categories, which are presented in Table 6.6. Among the ice classes, PC 1 is the strongest and PC7 is the weakest. PC6 and PC7 are close to Finnish-Swedish ice classes IA Super and IA, see Section 6.2.4.

6.2.3.2 Ice-Strengthening Principles of IACS PC Rules for the Hull

The hull of Polar Class ships is divided into areas reflecting the magnitude of the loads that are expected to act upon them. In the longitudinal direction, there are four regions: bow, bow intermediate, midbody and stern. The bow intermediate, midbody and stern regions are further divided in the vertical direction into the bottom, lower and icebelt regions. The extent of each hull area is illustrated in Figure 6.10 and the used hull shape parameters in Figure 6.11.

Design ice forces calculated according to the formulations given below are applicable for bow forms where the buttock angle γ at the stem is positive and less than 80°, and the normal frame angle β' at the centre of the foremost sub-region is greater than 10°. Separate requirements have been specified for bulbous ships.

TABLE 6.6
Definition of PC Classes (IACS, 2019)

Polar Class	Ice Descriptions (Based on WMO Sea Ice Nomenclature)
PC 1	Year-round operation in all polar waters
PC 2	Year-round operation in moderate multiyear ice conditions
PC 3	Year-round operation in second-year ice which may include multiyear ice inclusions.
PC 4	Year-round operation in thick first-year ice which may include old ice inclusions
PC 5	Year-round operation in medium first-year ice which may include old ice inclusions
PC 6	Summer/autumn operation in medium first-year ice which may include old ice inclusions
PC 7	Summer/autumn operation in thin first-year ice which may include old ice inclusions

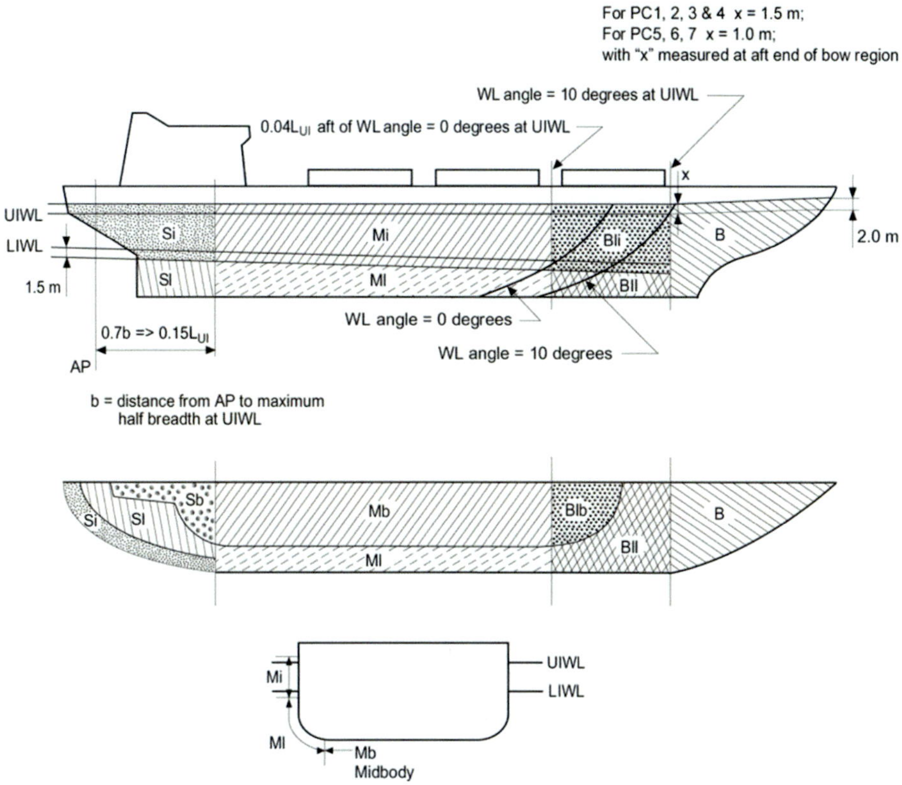

FIGURE 6.10 The hull regions defined in the IACS PC rules (IACS, 2019).

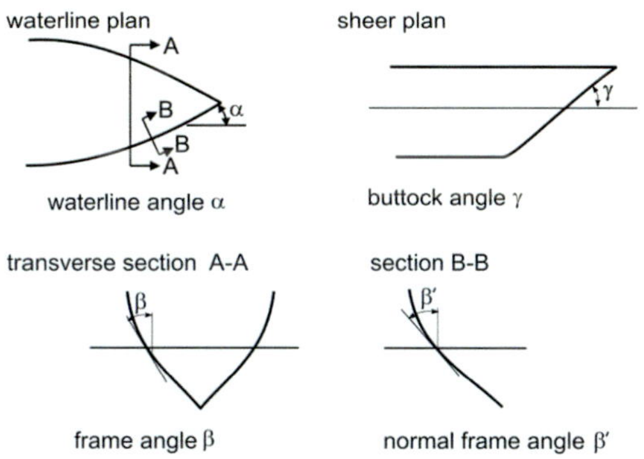

Note: β' = normal frame angle at upper ice waterline [deg]
α = upper ice waterline angle [deg]
γ = buttock angle at upper ice waterline (angle of buttock line measured from horizontal) [deg]
$\tan(\beta) = \tan(\alpha)/\tan(\gamma)$
$\tan(\beta') = \tan(\beta)\cdot\cos(\alpha)$

FIGURE 6.11 The hull shape parameters in the IACS PC rules (IACS, 2019).

A glancing impact on the bow is the design scenario for determining the scantlings required to resist ice loads (Daley, 1999). The design ice load is characterized by an average pressure (P_{avg}) uniformly distributed over a rectangular load patch of height (b) and width (w). Within the bow area of all Polar Class ships, and within the bow intermediate icebelt area of Polar Class PC6 and PC7, the ice-load parameters are functions of the actual bow shape. To determine the ice-load parameters (P_{avg}, b and w), it is required to calculate the following ice-load characteristics for sub-regions of the bow area: shape coefficient (fa_i), total glancing impact force (F_i), line load (Q_i) and pressure (P_i).

For bow area, the used formulations are:

1. Shape coefficient, fa_i, is to be taken as:

$$fa_i = \text{minimum}\left(fa_{i,1}; fa_{i,2}; fa_{i,3}\right)$$

where

$$fa_{i,1} = \left(0.097 - 0.68\left(\left(\frac{x}{L_{UI}} - 0.15\right)^2\right)\alpha_i/(\beta_i^{\prime})^{0.5}\right) \tag{6.17}$$

$$fa_{i,2} = 1.2CF_F/(\sin(\beta_i^{\prime})CF_c D_{UI}^{0.64}) \tag{6.18}$$

$$fa_{i,3} = 0.60 \tag{6.19}$$

2. Force, F_i:

$$F_i = fa_i CF_c D_{UI}^{0.64} \tag{6.20}$$

3. Load patch aspect ratio, AR_i:

$$AR_i = 7.46\sin(\beta_i^{\prime}) \geq 1.3 \tag{6.21}$$

4. Line load, Q_i:

$$Q_i = \frac{F_i^{0.61} CF_D}{AR_i^{0.35}}\left(\frac{\text{MN}}{\text{m}}\right) \tag{6.22}$$

5. Pressure, P_i:

$$P_i = F_i^{0.22} CF_D^2 AR_i^{0.3}\,(\text{MPa}) \tag{6.23}$$

where

i = sub-region considered;

L_{UI} is the distance, in m, measured horizontally from the foreside of the stem at the intersection with the upper ice waterline (UIWL) to the afterside of the rudder (m);

x is a distance from the foreside of the stem at the intersection with the upper ice waterline (UIWL) to station under consideration (m);
α is waterline angle [deg], see Figure 6.11;
β' is normal frame angle [deg], see Figure 6.11;
D_{UI} is the ship displacement, in kt, of the ship corresponding to the upper ice waterline (UIWL), not to be taken less than 5 (Katona);
CF_C is crushing failure class factor, varying as a function of ice class (PC1:17.69, PC7:1.8);
CF_F is flexural failure class factor varying as a function of ice class (PC1:68.60, PC7:4.06);
CF_D is load patch dimensions class factor varying as a function of ice class (PC1:2.01, PC7:1.11);

Design load patch for the bow area has dimensions of width, w_{Bow}, and height, b_{Bow}, defined as follows:

$$w_{Bow} = \frac{F_{Bow}}{Q_{Bow}} \text{(m)} \tag{6.24}$$

$$b_{Bow} = \frac{Q_{Bow}}{P_{Bow}} \text{(m)} \tag{6.25}$$

In the hull areas other than the bow, the force (F_{NonBow}) and line load (Q_{NonBow}) used in the determination of the load patch dimensions (b_{NonBow}, w_{NonBow}) and design pressure (P_{avg}) are determined by separate equations, being somewhat simpler as no hull shape parameters are needed.

The average pressure, P_{avg}, within a design load patch is determined as follows:

$$P_{avg} = \frac{F}{(bw)} \text{(MPa)} \tag{6.26}$$

Areas of higher, concentrated pressure exist within the load patch. In general, smaller areas have higher local pressures. Accordingly, the peak pressure factors listed in Table 6.7 are used to account for the pressure concentration on localized structural members.

Associated with each hull area is an area factor that reflects the relative magnitude of the load expected in that area. The Area Factor (AF) for each hull area is listed and they vary as a function of area and ice class (varying from 1.0 to 0.25). Due to their increased manoeuvrability, ships having propulsion arrangements with azimuth thruster(s) or 'podded' propellers are to have specially considered Stern Icebelt (Si) and Stern Lower (Sl) hull area factors.

Once the load patch is determined the required plate thickness can be determined from:

$$t = t_{net} + t_s \tag{6.27}$$

where t_{net} is plate thickness required to resist ice loads accordingly, see below.

TABLE 6.7
Definition of Peak Pressure Factors (IACS, 2019)

Structural Member		Peak Pressure Factor (PPF_i)
Plating	Transversely-framed	$PPF_p = (1.8 - s) \geq 1.2$
	Longitudinally-framed	$PPF_p = (2.2 - 1.2 \cdot s) \geq 1.5$
Frames in transverse framing systems	With load distributing stringers	$PPF_t = (1.6 - s) \geq 1.0$
	With no load distributing stringers	$PPF_t = (1.8 - s) \geq 1.2$
Frames in bottom structures		$PPF_s = 1.0$
Load-carrying stringers		$PPF_s = 1.0$, if $S_w \geq 0.5$ w
Side longitudinals		$PPF_s = 2.0 - 2.0\ S_w/w$, if $S_w < (0.5 \cdot w)$
Web-frames		

Where: s = frame or longitudinal spacing (m); S_w = web-frame spacing (m); w = ice load patch width (m).

TABLE 6.8
Definition of Corrosion/Abrasion Additions for Shell Plating (IACS, 2019)

	t_s [mm]					
	With Effective Protection			Without Effective Protection		
Hull Area	PC1–PC3	PC4 & PC5	PC6 & PC7	PC1–PC3	PC4 & PC5	PC6 & PC7
Bow; Bow Intermediate Icebelt	3.5	2.5	2.0	7.0	5.0	4.0
Bow Intermediate Lower; Midbody & Stern Icebelt	2.5	2.0	2.0	5.0	4.0	3.0
Midbody & Stern Lower; Bottom	2.0	2.0	2.0	4.0	3.0	2.5

t_s is corrosion and abrasion allowance. Effective protection against corrosion and ice-induced abrasion is recommended for all external surfaces of the shell plating for Polar Class ships. The values of corrosion/abrasion additions, t_s, to be used in determining the shell plate thickness are listed in Table 6.8. Polar Class ships are to have a minimum corrosion/abrasion addition of $t_s = 1.0$ mm applied to all internal structures within the ice-strengthened hull areas, including plated members adjacent to the shell, as well as stiffener webs and flanges.

In case of transversely framed plating, the net thickness is given by:

$$t_{net} = 500s\sqrt{\left(\frac{AF*PPF_p*P_{avg}}{\sigma_y}\right)}/(1 + s/(2*l)) \tag{6.28}$$

In case of longitudinally framed plating, the net thickness is given by:

$$t_{net} = 500s\sqrt{\left(\frac{AF*PPF_p*P_{avg}}{\sigma_y}\right)}*\sqrt{\left(2*\frac{b}{s}-\left(\frac{b}{s}\right)^2\right)}/(1+s/(2*l)) \qquad (6.29)$$

The local frames in bottom structures and transverse local frames in side structures are to be dimensioned such that the combined effects of shear and bending do not exceed the plastic strength of the member. The plastic strength is defined by the magnitude of midspan load that causes the development of a plastic collapse mechanism. The rules give detailed instruction on how to determine the plastic strength.

6.2.4 Comparison on Various Ice Classes

As stated above, different ice-class rules have totally different formulations to determine the ice-induced loads and related scantlings of the hull. All rules have ship displacement as the main parameter to determine the load level. This is natural as the kinetic energy of the ships is important physical parameter on the ship-ice interaction, see also Chapter 4. Only FSICR has ship engine power as one parameter to define the load level. Hull shape has important role in Russian and IACS PC rules, but hull shape is not included on the hull ice-load formulations of FSICR. While Russian and IACS PC rules use the ultimate strength of frames and other girder systems as the strength criterion, FSICR adopts first yield as the basis for strength criterion. This means that the load level used on various rules cannot be compared as the load formulations is also related to the used strength criteria, see Section 6.5.

All the classification societies and Arctic states administrations have developed equivalent tables to compare the various ice classes described above. Appolonov et al. (2007) made an interesting analysis, where they compared the used design ice thickness and obtained cross-sectional area of the ice-strengthening region (i.e., this has clear link to hull weight) as shown in Figures 6.12 and 6.13.

6.3 PROPULSION RULES

The earlier ice rules formulations were based on Ignatjev (1964) ice moment approach. For Finnish-Swedish Ice-Class Rules (FSICR), it was further developed by Enkvist and Johansson (1968). In 1971, FSICR rules, the ice torque for the propellers, propeller shaft and gearing were based on the formula

$$M = mD^2\left(t\cdot m^2\right) \qquad (6.29)$$

where D is diameter of the propeller in meters and m is a constant varying from 2.15 (IAS) to 1.22 (IC). In addition, a simple formula is given to determine propeller width and plate thickness on various radius distances and the affecting parameters used in the rule formulations were ice torque, propeller diameter, shaft horsepower, propeller revolutions, number of blades, propeller pitch and tensile strength of the used material.

The current ice-class rule approach for the Polar Code and FSICR is based on the extensive research done at VTT and Helsinki University of Technology as described

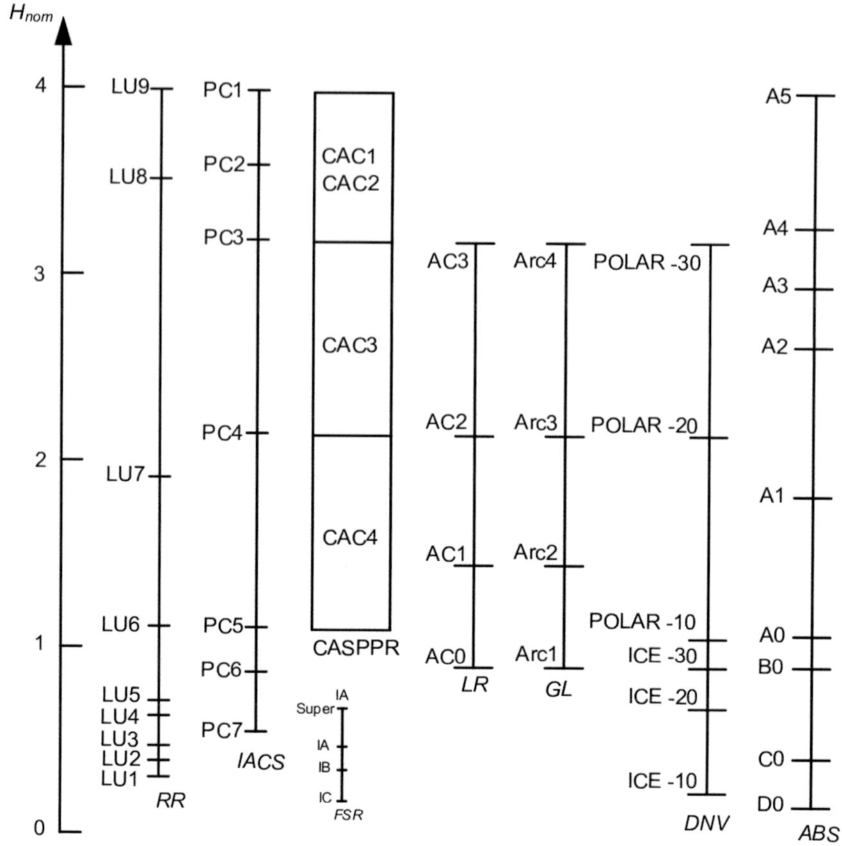

FIGURE 6.12 Comparison of various ice-class rules design ice thickness.RR – Russian Register; IACS PC – Unified Requirements of International Association Classification Societies; CASPPR – Canadian Arctic Shipping Pollution Prevention Regulation; FSR – Finnish-Sweden Rules; LR – Lloyd's Register; GL – Germanisher Lloyd; DNV – Det Norske Veritas; ABS – American Bureau of Shipping.

(Re-produced from Appolonov et al., 2007.)

by Veitch (1995), Koskinen et al. (1996) and Soininen (1998) and summarized by IACS (2006) on the background notes for rule development. The current IACS rules define formulations to determine minimum blade force when bending the blade backwards and forward, blade bending moment, spindle torque, shaft thrust and torque and tip load. Separate guidance is given for ducted propellers and highly skewed propellers. Full-scale ice-load data from the following ships is mentioned in Koskinen et al. (1996): MS Gudingen, IB Lemeur, IB Sampo and IB Karhu. They are used to calibrate the used theoretical models.

The propeller ice-load calculation method as per Finnish-Swedish ice-class rules and IACS (2006) are hereby introduced based on the report by Koskinen et al. (1996). The following notations are applied: F_b is the maximum backward-bending force on the propeller blade when the propeller is turned forward to strike ice during the life of

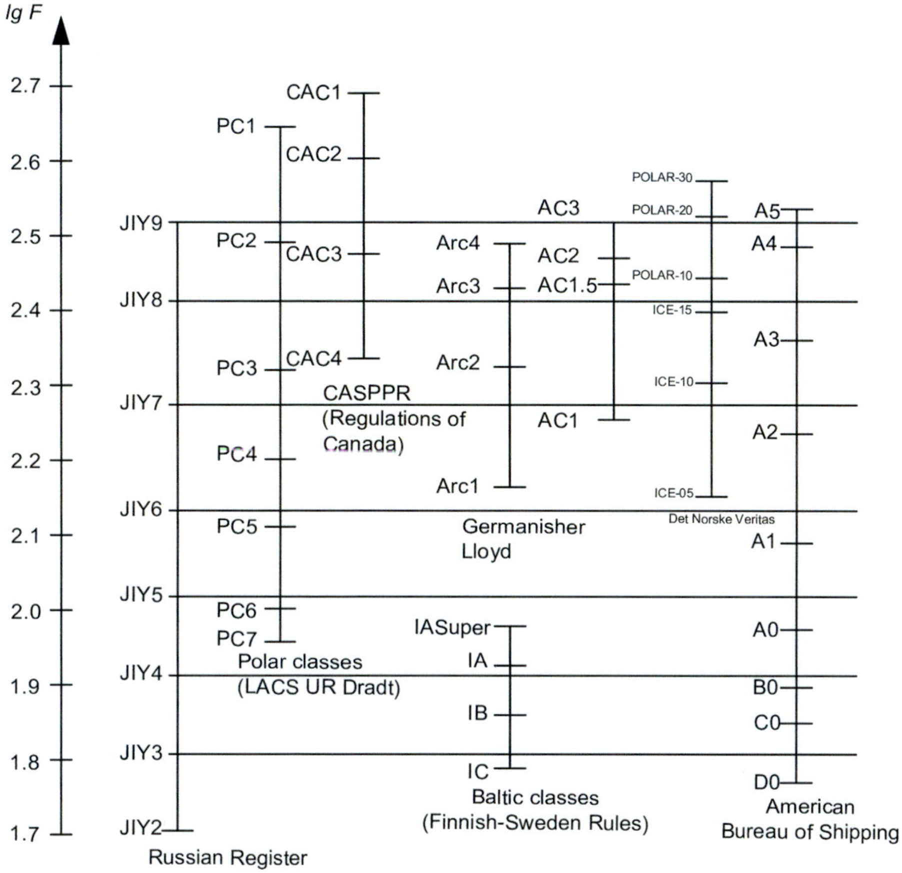

FIGURE 6.13 Comparison of various ice-class cross-sectional area.

(Re-produced Appolonov et al., 2007.)

the ship; F_f is the maximum forward-bending force on the propeller blade when the propeller is turned forward to strike ice during the life of the ship. These forces arise from different phenomena of propeller-ice interaction and are not simultaneous, so they should be applied to a blade separately.

1. Maximum backward force on propeller blade F_b
   when $D \leq D_{\text{limit}}$,

$$F_b = 27S_{ice}(nD)^{0.7}\left(\frac{EAR}{Z}\right)^{0.3}D^2$$

   when $D > D_{\text{limit}}$,

$$F_b = 23S_{ice}(nD)^{0.7}\left(\frac{EAR}{Z}\right)^{0.3}DH_{ice}^{1.4}$$

TABLE 6.9
Definition of Design ice Thickness and Ice Strength Index for Various Ice Classes (IACS, 2006)

Ice Class	H_{ice} [m]	S_{ice} [-]	S_{qice} [-]
PC1	4.0	1.2	1.15
PC2	3.5	1.1	1.15
PC3	3.0	1.1	1.15
PC4	2.5	1.1	1.15
PC5	2.0	1.1	1.15
PC6	1.75	1	1
PC7	1.5	1	1

where $D_{\text{limit}} = 0.85\, H_{ice}^{1.4}$, in meters; for controllable pitch propellers, n is the calibrated RPM at maximum continuous power free running without ice; for fixed pitch propellers, n is 85% of the calibrated RPM at maximum continuous power free running without ice. D is propeller diameter in metres, Z is number of blades, EAR is expanded blade area ratio, S_{ice} is ice strength index, see Table 6.9.

2. The forward blade load on the open propeller is a non-contact load occurring due to the close proximity of an ice block and propeller blade. These loads have been measured directly in full scale, but the exact mechanism of their generation and the shape of the load distribution on the blade are not fully understood. Maximum forward force on propeller blade F_f.

When $D \leq D_{\text{limit}}$,

$$F_f = 250 \frac{EAR}{Z} D^2$$

When $D > D_{\text{limit}}$,

$$F_f = 500 \frac{EAR}{Z} D \left(\frac{1}{1 - \frac{d}{D}} \right) H_{ice}^{1.4}$$

where $D_{limit} = \left(\frac{2}{1 - \frac{d}{D}} \right) H_{ice}$, in meters and d/D is the hub ratio.

Controllable and fixed-pitch propellers should consider load conditions 1–4 shown in Table 6.10. Fixed-pitch propellers should also consider load condition 5 in order to reflect blade ice loading when the propeller is inverted. The designers can use these calculation methods to determine the ice loads and analyse the strength of the propeller.

The above regulation gives instructions on how to estimate the extreme ice load on the blade. During the operation of ships in ice, the propeller receives continuous, cyclic loading from the impact with ice, which may incur fatigue damage. IACS (2006) gives instructions on how to obtain the load spectrum for the analysis of fatigue.

TABLE 6.10
Propeller Load Condition (IACS, 2006)

Load Condition	Force	Stress area	Clockwise Rotating Propeller Blades Viewed from the Stern of the Boat
1	F_b	A force applied uniformly on the blade suction surface over a region extending from 0.6 R to the tip of the blade and extending 0.2 times the chord length from the leading edge.	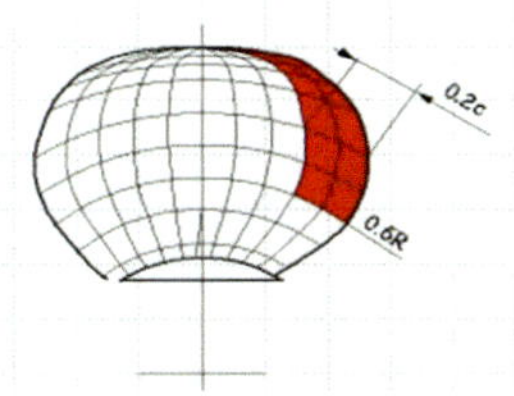
2	$50\%F_b$	Pressure uniformly applied on the blade suction surface to the tip region of the propeller beyond 0.9 R.	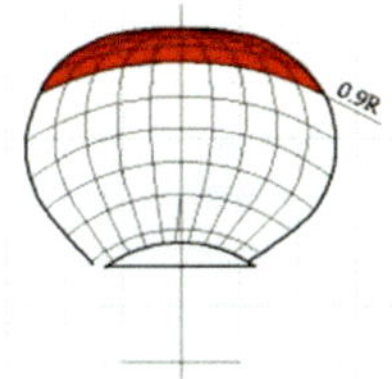
3	F_f	Pressure uniformly applied on the blade pressure surface over a region extending from 0.6 R to the tip of the blade and extending 0.2 times the chord length from the leading edge.	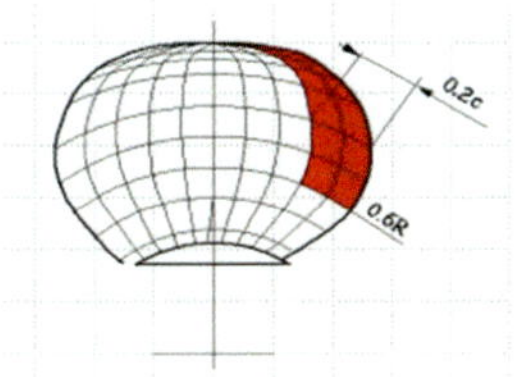
4	$50\%F_f$	Pressure uniformly applied on the blade pressure surface to the tip region of the propeller beyond 0.9 R.	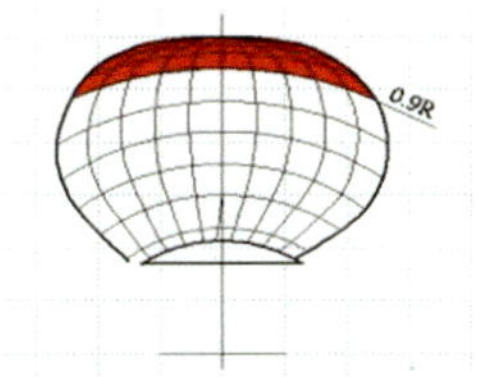
5	$60\%F_f$ or $60\%F_b$, whichever is greater	Pressure applied uniformly on the blade pressure surface over a region extending from 0.6 R to the tip of the blade and from the following edge to 0.2 times the chord length.	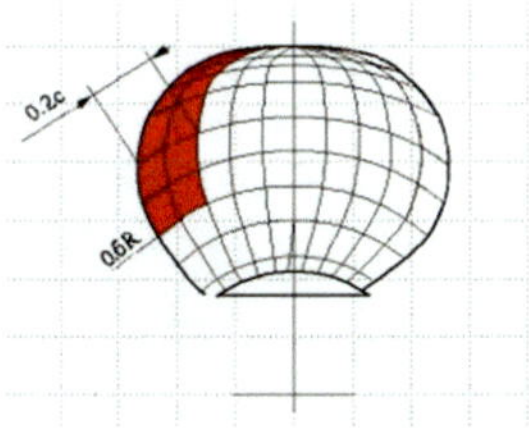

3. The formula for maximum propeller ice torque (ice torque or polar moment applied to the shaft by the propeller at the propeller location) is described as (IACS, 2006):

$$D_{\text{limit}} = 1.81H_{ice}$$

When $D < D_{\text{limit}}$:

$$Q_{max} = 105\left(1 - \frac{d}{D}\right)S_{qice}\left(\frac{P_{0.7}}{D}\right)^{0.16}\left(\frac{t_{0.7}}{D}\right)^{0.6}(nD)^{0.17}D^3 \text{ kNm}$$

When $D > D_{\text{limit}}$:

$$Q_{max} = 202\left(1 - \frac{d}{D}\right)S_{qice}H_{ice}^{1.1}\left(\frac{P_{0.7}}{D}\right)^{0.16}\left(\frac{t_{0.7}}{D}\right)^{0.6}(nD)^{0.17}D^{1.9} \text{ kNm}$$

where P is pitch and t is blade thickness and they are taken at 0.7 radius in m. A coefficient S_{qice} to account for the ice strength influence, see Table 6.8.

Figures 6.14 and 6.15 illustrate few examples of the load obtained by the equations above for backward and forward ice load on the propeller blade. The maximum

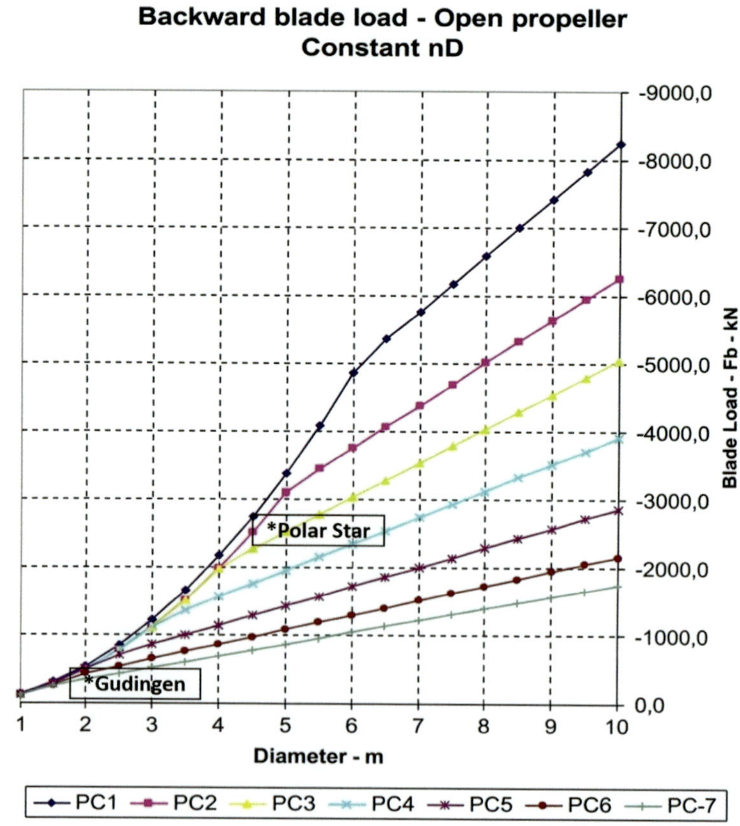

FIGURE 6.14 Backward blade load with $nD = 12$, $EAR = 0.7$ and $Z = 4$ (IACS, 2006).

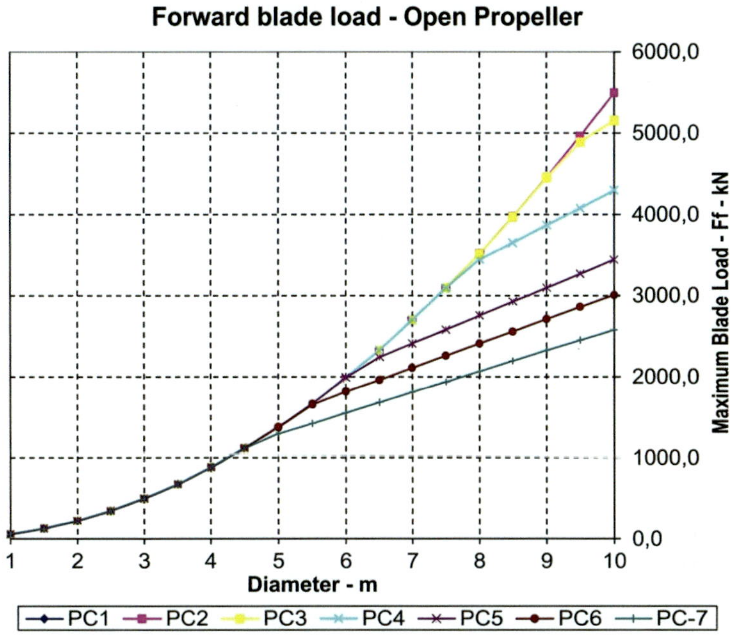

FIGURE 6.15 Forward blade load with $d/D = 0.36$, $nD = 12$, $EAR = 0.7$ and $Z = 4$ (IACS, 2006).

directly measured blade loads for the vessels *Gudingen* – PC7 equivalent and *Polar Star* – PC3 equivalent are also shown in Figure 6.14 to demonstrate the overall validity of the load models.

6.4 OPERATIONAL RULES, POLAR CODE

6.4.1 BACKGROUND

The International Maritime Organization (IMO) has developed the mandatory International Code of Safety for Ships Operating in Polar Waters (IMO, Polar Code, 2017).

The Polar Code has been developed to supplement the existing IMO instruments in order to improve the safety of shipping and to mitigate harmful effects of shipping on the environment in the remote, vulnerable and potentially harsh polar waters, see Hindley et al. (2023) to understand the main aims of the code.

The code covers a full range of design, construction, equipment, operational, training, search and rescue and environmental protection matters relevant to ships operating in the inhospitable waters surrounding the two poles, see Figure 6.16. The Polar Code came into force in the beginning of 2017.

The two main conventions of the International Maritime Organization (IMO) related to safety and pollution prevention from ships are:

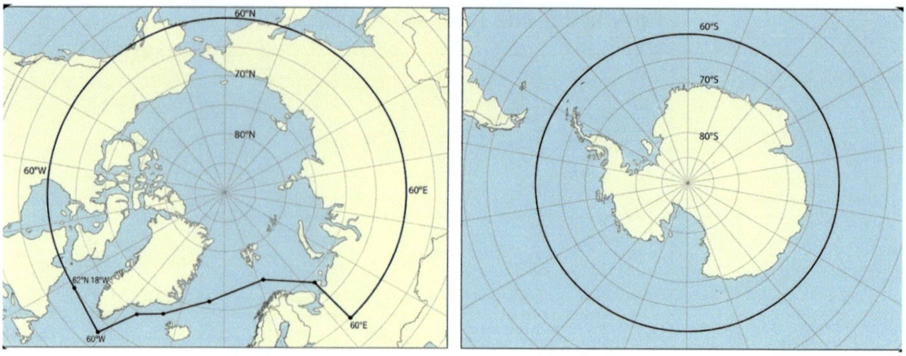

FIGURE 6.16 Definition of Arctic and Antarctica sea area for Polar Code (2017).

International Convention for the Safety of Life at Sea (SOLAS [IMO, 1974]); and

International Convention for the Prevention of Pollution from Ships (MARPOL 73/78 [IMO, 2002]).

The Polar Code will be adopted by adding a new chapter: Safety Measures for Ships Operating in Polar Waters, to the SOLAS Convention, and by adopting relevant amendments to the MARPOL Convention.

6.4.2 Sources of Hazards Leading to Elevated Levels of Risk in Polar Waters

The Polar Code considers hazards which may lead to elevated levels of risk due to increased probability of occurrence, more severe consequences, or both:

1. ice,
2. topside icing,
3. low temperature,
4. extended periods of darkness or daylight,
5. high latitude,
6. remoteness and possible lack of accurate and complete hydrographic data and information,
7. potential lack of ship crew experience in polar operations,
8. potential lack of suitable emergency response equipment,
9. rapidly changing and severe weather conditions,
10. the environment with respect to sensitivity to harmful substances and other environmental impacts and its need for longer restoration.

6.4.3 Content

The Polar Code includes mandatory measures covering safety part (part I-A) and pollution prevention (part II-A) and recommendatory provisions for both (parts I-B and II-B):

POLAR CODE - STRUCTURE
Part I-A – Safety measures (mandatory)
(12 chapters – Largely goal-based)

1. General
2. Polar water operational manual
3. Ship structure
4. Subdivision and stability
5. Watertight and weathertight integrity
6. Machinery installations
7. Fire safety/protection
8. Life-saving appliances and arrangements
9. Safety of navigation
10. Communication
11. Voyage planning
12. Manning and training

Part I-B – Pollution prevention measures (mandatory)
(4 chapters – Prescriptive)

1. Prevention of pollution by oil
2. Control of pollution by noxious liquid
3. Prevention of pollution by harmful substances carried by sea in packaged form
4. Prevention of pollution by sewage from ships

The code will require ships intending to operate in the defined waters of the Antarctic and Arctic to apply for a Polar Ship Certificate, which would classify the vessel as Category A ship – ships designed for operation in polar waters at least in medium first-year ice, which may include old ice inclusions. These are typically ships with IACS ice-classes PC1 to PC 5, see Section 6.2.3

Category B ship – a ship not included in category A, designed for operation in polar waters in at least thin first-year ice, which may include old ice inclusions. These are typically ships with IACS ice-classes PC6 to PC 7, see Section 6.2.3.

Category C ship – a ship designed to operate in open water or in ice conditions less severe than those included in Categories A and B. These are typically ships with FSICR ice-classes IA Super and IA, see Section 6.2.1.

The issuance of a certificate would require an assessment, taking into account the anticipated range of operating conditions and hazards the ship may encounter in the polar waters. The assessment would include information on identified operational limitations, and plans or procedures or additional safety equipment necessary to mitigate incidents with potential safety or environmental consequences.

Ships need to carry a Polar Water Operational Manual, to provide the owner, operator, master and crew with sufficient information regarding the ship's operational capabilities and limitations in order to support their decision-making process.

The chapters in code set out goals and functional requirements, to include those covering ship structure, stability and subdivision, watertight and weathertight

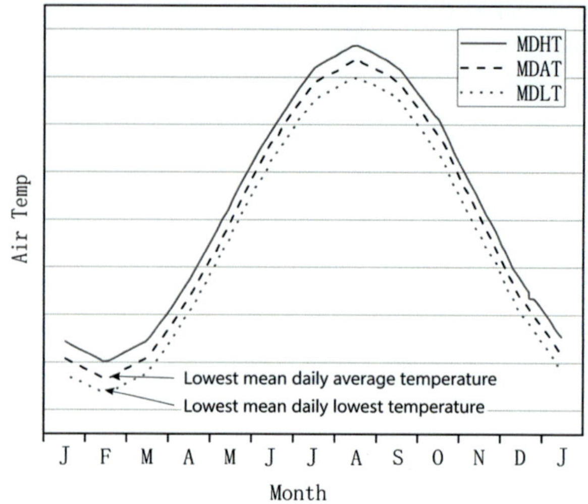

Definitions used in the figure above

MDHT–Mean Daily High Temperature

MDAT–Mean Daily Average Temperature

MDLT–Mean Daily Low Temperature

FIGURE 6.17 Definition of design temperature in Polar Code.

integrity, machinery installations, operational safety, fire safety/protection, life-saving appliances and arrangements, safety of navigation, communications, voyage planning, manning and training, prevention of oil pollution, prevention of pollution form from noxious liquid substances from ships, prevention of pollution by sewage from ships, and prevention of pollution by discharge of garbage from ships.

One new element of the Polar Code is the definition of a design temperature to be used in selecting materials for the ship hull and for equipment installed on the deck. Polar Service Temperature (PST) means a temperature specified for a ship which is intended to operate in low air temperature, which shall be set at least 10°C below the lowest MDLT for the intended area and season of operation in polar waters. Mean Daily Low Temperature (MDLT) is the mean value of the daily low temperature over a minimum ten-year period (see Figure 6.17).

6.4.4 POLARIS

Polar Operational Limit Assessment Risk Indexing System (POLARIS) is a standard approach for determining the limitations in operations in ice; it is not obligatory. It can be used for voyage planning purposes or in real time to aid in decision-making by administrations as a means to set operational limitations for ships. POLARIS uses a Risk Index Outcome (RIO) value to assess limitations for operation in ice (MSC. 1/Circ 1519, 2016). Risk Index Values (RIVs) are assigned to the ship based on ice class. For each ice regime encountered, the Risk Index Values are used to determine a RIO that forms the basis of the decision to operate or limitation for operation. The RIO is determined by a summation of the RIVs for each ice type present in the ice regime multiplied by its concentration (expressed in tenths):

$$\mathrm{RIO} = (C_1\mathrm{RIV}_1) + (C_2\mathrm{RIV}_2) + (C_n\mathrm{RIV}_n) \tag{6.30}$$

where $C_1 \ldots C_n$ are the concentrations (in tenths) of ice types within the ice regime and $\mathrm{RIV}_1 \ldots \mathrm{RIV}_n$ are the Risk Index values corresponding to the ships ice class, Table 6.11.

The operational limitations for ships operating independently are based on the evaluation criteria so that the RIO should be positive to indicate normal operations and when the value of RIO is below 0 but higher than -10, limited speed requirements is established for various ice classes of PC1 to PC7, for other ice-classes, operations in ice should be avoided for voyage planning purposes when RIO is negative. For ships operating under icebreaker escort, where the track created by the icebreaker(s) is wider than the beam of the escorted ship, the operational limitations are primarily determined by the ice conditions within the channel behind the icebreaker. For voyage planning purposes when ice-breaker is intended to be used, the RIO derived from non-escorted historical ice data may be assumed to be modified by adding 10 to its calculated value, given that due caution of the mariner will be exercised, taking into account such factors as changes in weather and visibility.

Table 6.12 shows the criteria for RIO values and when RIO value is between 0 and -10, ship should deduce the speed as shown in Table 6.13.

6.5 RISK-BASED DESIGN

As stated above, conventional rule-based ship design for ice conditions include a number of semi-empirical equations. Although rule or prescriptive-based design approaches are easy to follow and implement, they may cause over-conservative design. Safety level is often uncertain or unknown in design rules and rules do not always reflect experience. Therefore, a move towards a goal or performance-based approach to ship design would be preferable over the conventional approaches. A risk-based design approach is one way forward in this direction (Kujala et al., 2019). This implies that a risk-based design can help to meet safety expectations in cost-effective ways.

The underlying rationale of such a proactive approach to safe ship design is in line with the philosophy adopted by the International Maritime Organization (IMO) to apply goal-based (opposed to rule-based) ship design standards (MSC.1/Circ.1394/Rev.1, 2015). The focus on risk-based approaches in international maritime regulatory decision-making is also clear from the Formal Safety Assessment (FSA) framework adopted by the IMO (MSC-MEPC.2/Circ.12/Rev.1, 2015). For the marine industry, risk assessment is nothing new. Classification societies have published general guidelines on risk assessment and its applications to ship design and operations (e.g., ABS, 2000; ABS, 2003).

TABLE 6.11
Risk Index Values (RIV) for Various Ice Classes, Winter Conditions

Ice Class	Ice Free	New Ice	Grey Ice	Grey White Ice	Thin First Year Ice, 1st Stage	Thin First Year Ice, 2nd Stage	Medium First Year Ice, less than 1 m thick	Medium First Year Ice	Thick First Year Ice	Second Year Ice	Light Multi Year Ice, less than 2.5 m thick	Heavy Multi Year Ice
PC1	3	3	3	3	2	2	2	2	2	2	1	1
PC2	3	3	3	3	2	2	2	2	2	1	1	0
PC3	3	3	3	3	2	2	2	2	2	1	0	−1
PC4	3	3	3	3	2	2	2	2	1	0	−1	−2
PC5	3	3	3	3	2	2	1	1	0	−1	−2	−2
PC6	3	2	2	2	2	1	1	0	−1	−2	−3	−3
PC7	3	2	2	2	1	1	0	−1	−2	−3	−3	−3
IA Super	3	2	2	2	2	1	0	−1	−2	−3	−4	−4
IA	3	2	2	2	1	0	−1	−2	−3	−4	−5	−5
IB	3	2	2	1	0	−1	−2	−3	−4	−5	−6	−6
IC	3	2	1	0	−1	−2	−3	−4	−5	−6	−7	−8
No ice class	3	1	0	−1	−2	−3	−4	−5	−6	−7	−8	−8

TABLE 6.12
RIO Criteria

RIO	Type of operation
RIO ≥ 0	Normal operation
–10 ≥ RIO < 0	Elevated operational risk
RIO < –10	Operation subject to special consideration

TABLE 6.13
Recommended Speed Limits

Ice class	Recommended speed limit
PC1	11 knots
PC2	8 knots
PC3–PC5	5 knots
Below PC5	3 knots

The evaluation of the failure probability of the ship hull is straightforward once the long-term load level is known. Typically, Gumbel 1 extreme value distribution can be fitted to the measured ice-load data to forecast the lifetime extreme loads acting on the ship. By comparing these load values with the ultimate strength of the frames using various ice classes to determine the scantlings, the probability of reaching the planned limit state can be evaluated. This has been done by Kujala (1991) and Kaldasaun and Kujala (2011). As a result, a reliability index β is obtained by defining a probability of failure for the shell structures of the ship. This approach is only possible when full-scale long-term measurement data are available, and it is only valid for the specific ship instrumented.

Riska (2011) has made an extensive study to evaluate the design point for ice-class vessel. The *MT Kemira* full-scale data is used as a basis for analysing the ice-class rules and also the design point. The main conclusion is that, on average, the yield strength is reached annually during navigation in ice, and the ultimate limit state is reached once over the ship's lifetime. This means that, for plating, allowing once-in-winter yield-level response results in approximately the same required plate thickness as allowing once-in-a-lifetime fully plastic response.

The knowledge of the variation of the contact on the ship hull and its relation to the ice-induced loads in various ice conditions and operations is still too limited to form a good basis for risk-based design and this is the main challenge to apply risk-based design for ice-going ships (Kujala et al., 2019, Bergström et al., 2022a).

Bergstrom et al. (2022b) made a comprehensive risk-based design using semi-empirical methods to evaluate the long-term ice loading on ships. The credibility of these methods comes from the fact that they merge full-scale ice-load measurements with the existing physical understanding of the ship-ice interaction process and the resulting ice loading (e.g., knowledge about the relationship between ice pressure and the ship-ice contact area). As a result, in comparison with purely empirical models, semi-empirical models are applicable to a wider range of design cases and compared with purely theoretical models; semi-empirical models can account for those ship-ice interaction process-related phenomena that are not fully known. Among the above-presented semi-empirical methods, for assessing ice load on ships operating in Arctic waters, Bergström et al. (2022b) believe that the modified event-maximum method is among the most appropriate readily available methods (see Figure 6.18). The utility of the proposed approach is demonstrated through a case study, in which it is used to select an appropriate Polar Class for a double-acting ship intended for year-round independent operations along the northeast coast of Canada. For the bow shoulder area, the calculated 25-year load was found to be close to the corresponding PC 3 design load (see Figure 6.19).

The few examples given above show clearly the potential of the risk-based design approach to optimize the future ship hull and propulsion concepts for icy waters. The main challenge is still how to relate the ice-induced loads to the prevailing ice

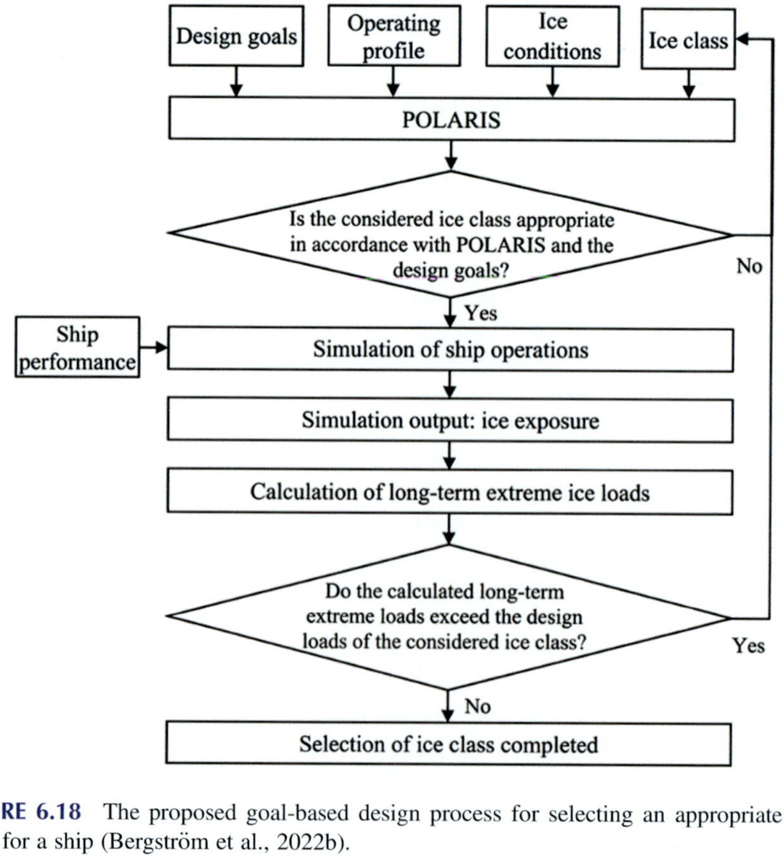

FIGURE 6.18 The proposed goal-based design process for selecting an appropriate Polar Class for a ship (Bergström et al., 2022b).

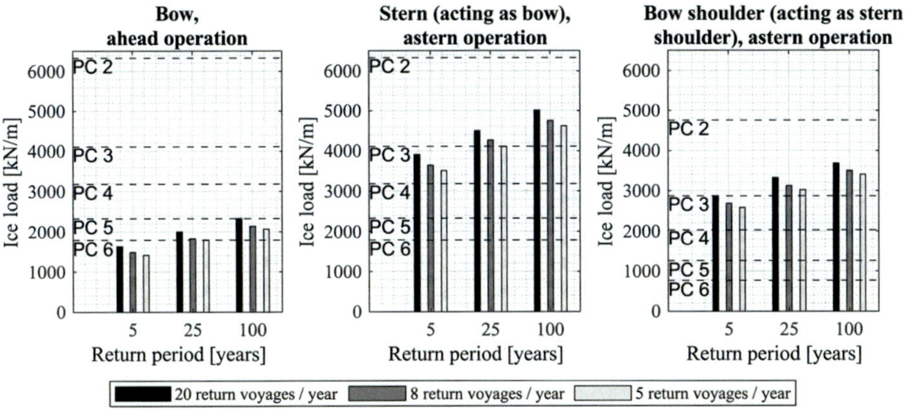

FIGURE 6.19 Calculated ice loads for different hull areas, return periods and operating profiles (Bergström et al., 2022b).

conditions and ship operation principles as discussed in Chapters 3 and 4. Once this is resolved, risk-based design will enable in future the best approach for designing ice-going ships under varying ice conditions.

6.6 SUMMARY

Different ice-class rules have totally different formulations to determine the ice-induced loads and related scantlings of the hull. All rules have ship displacement as the main parameter to determine the load level. This is natural as the kinetic energy of the ships is important physical parameter on the ship-ice interaction. Only FSICR uses ship engine power as a parameter to define the load level. Hull shape has an important role to play in Russian and IACS PC rules, but hull shape is not included in the hull ice-load formulations of FSICR. The Russian and IACS PC rules are based on ultimate strength of frames and other girder systems, while FSICR is based on first yield as the strength criterion. This means that the load level used on various rules cannot be comparted as the load formulations is also related to the used strength criteria.

The International Maritime Organization (IMO) has developed the mandatory International Code of Safety for Ships Operating in Polar Waters. The Polar Code has been developed to supplement the existing IMO instruments in order to improve the safety of shipping and to mitigate harmful effects of shipping on the environment in the remote, vulnerable and potentially harsh polar waters. The code covers a full range of design, construction, equipment, operational, training, search and rescue and environmental protection matters relevant to ships operating in the inhospitable waters surrounding the two poles.

Conventional rule-based ship design for ice conditions includes a number of semi-empirical equations. Although rule or prescriptive-based design approaches are easy to follow and implement, they may cause over-conservative design. Safety level is often uncertain or unknown in design rules and rules do not always reflect experience. Therefore, a move towards a goal or performance-based approach to ship design would be preferable over the conventional approaches. A risk-based design approach is one way forward in this direction. This implies that a risk-based design can help to meet safety expectations in cost-effective ways.

REFERENCES

ABS, 2000. *Risk assessment applications for marine and offshore oil and gas industries.* Houston, USA: ABS.

ABS, 2003. *Guidance notes on review and approval of novel concepts.* Houston, USA: ABS.

Appolonov, E. M., Nesterov, A. B., Paliy, O. M., and Timofeev, O. Y., 2007. A system of forming fundamental engineering solutions on assurance of ice strength and safe ship service in Russian arctic and freezing seas. In *Arctic Shipping Forum Conference*, Helsinki, April 2007.

Bergström, M., Browne, T., Ehlers, S., Helle, I., Herrnring, H., Khan, F., Kubiczek, J., Kujala, P., Korgesaar, M., Leira, B. J., Parviainen, T., Polojarvi, A., Suominen, M., Taylor, R., Tuhkuri, J., Vanhatalo, J., and Veitch, B., 2022b. A comprehensive approach to scenario-based risk management for Arctic waters. *Ship Technology Research.* 69(3), 129–157.

Bergström, M., Li, F., Suominen, M., and Kujala, P., 2022a. A goal-based approach for selecting a ship's polar class. *Marine Structures*, 81, 103123.

Daley, C., 1999, May. Energy based ice collision forces. In *Proceedings of the 15th International Conference on Port and Ocean Engineering under Arctic Conditions (POAC)*, Vol. 2, pp. 674–686, Espoo, Finland, 23–27 August.

Daley, C., 2014. *Ice class rules–Description and comparison.* Lecture notes. Memorial University of Newfoundland.

Enkvist, E., and Johansson, B. M., 1968. On icebreaker screw design. *European Shipbuilding*, 17(1), 1968.

FSICR, 1971. *Finnish-Swedish ice class rules 1971.* Helsinki: Board of Navigation, 43 p.

FSICR, 1985. *Finnish-Swedish ice class rules 1985.* Helsinki: Board of Navigation, 43 p.

Hindley, R., Kujala, P., and Hirdaris, S., 2023. Implementation of the polar code - Perspectives on the code five years on. *The 27th International Conference on Port and Ocean Engineering under Arctic Conditions*, Glasgow, UK, 12–14 June.

IACS Unified Requirement URI, 2006. *Structural requirements for polar class ships.* IACS Unified Requirement URI.

IACS Unified Requirement URI, 2019. *Structural requirements for polar class ships.* IACS Unified Requirement URI.

Ignatjev, M. A., 1964. Determination of ice loads encountered by ship propeller blades, (translated from Russian). *Problems of the Arctic and Antarctic*, 15, 41–51.

IMO, Polar Code, 2017. *International code for ships operating in polar waters (polar code).* IMO.

International Maritime Organisation (IMO), 1974. International convention for the safety of life at sea (SOLAS).

International Maritime Organisation (IMO), 2002. MARPOL 73/78: The international convention for the prevention of marine pollution from ships, 1973 as modified by its protocol of 1978.

Johansson, B., 1967. On the ice-strengthening of ship hulls. *International Shipbuilding Progress*, 14, 231–245.

Jones, N., 1976. Plastic behaviour of ship structures. *SNAME transactions*, 84, 115–145.
Juva, M., and Riska, K., 2003. On the power requirements in the Finnish-Swedish ice class rules. Winter Navigation Research Board. Report No. 52.
Kaldasaun, J., and Kujala, P., 2011. Risk-based approach for structural design of ice-strengthened vessels navigating in the Baltic Sea. In *Proceedings of the 21st International Conference on Port and Ocean Engineering under Arctic Conditions*, July 10–14, 2011, Montréal, Canada.
Körgesaar, M., Kujala, P., and Romanoff, J., 2018. Load carrying capacity of ice-strengthened frames under idealized ice load and boundary conditions. *Marine Structures*, 58, 18–30.
Koskinen, P., Jussila, M., Soininen, H., 1996. *Propeller ice load models*. Technical Research Centre of Finland, Research notes 1739.
Kujala, P., 1991. Safety of ice-strengthened ships in the Baltic Sea. London. Royal Institution of Naval Architects. *Transactions*.
Kujala, P., Goerlandt, F., Way, B., Smith, D., Yang, M., Khan, F., and Veitch, B., 2019. Review of risk-based design for ice-class ships. *Marine Structures*, 63, 181–195.
Law, 2005. Law of ship's ice class and icebreaker assistance. Valtion säädöstietopankki. 2005. (in Finnish).
MSC. 1/Circ 1519, 6 June 2016. *Guidance on methdodologies for assessing operational capabilities and limitations in ice*. IMO.
MSC-MEPC.2/Circ. 12/Rev.1, 18 June 2015. *Revised guidelines for formal safety assessment (FSA) for use in the IMO rule-making process*. IMO.
Popov, Y., Faddeyev, O., Kheisin, D., and Yalovlev, A., 1967. *Strength of ships sailing in ice*. Leningrad: Sudostroenie Publishing House.
Riska, K., 2011. Design point in ice class rules. Winter Navigation Research Board. Research Report No. 71. 40 p.
Riska, K., and Kämäräinen, J., 2011. A review of ice loading and the evolution of the finnish-swedish ice class rules. In *Proceedings of the SNAME Annual Meeting and Expo*, Houston, TX, 16–18 November.
Riska, K., Wilhelmsson, M., Englund, K., and Leiviskä, T., 1997. Performance of merchant vessels in ice in the Baltic. Winter Navigation Research Board. Report No. 52.
RMRS, 2010. Rules for the classification and construction of sea going ships. St. Petersburg.
RMRS, 2014. Rules for the classification and construction of sea going ships. St. Petersburg.
Soininen, H., 1998. *A propeller-ice contact model. dissertation for the degree of doctor of technology*. Espoo: Helsinki University of Technology, pp. 113–121.
Veitch B., 1995. Predictions of ice contact forces on a marine screw-propeller during the propeller-ice cutting process. *Acta Polytechnica Scandinavica Mechanical Engineering*, Helsinki, p. 140.

7 Experimental Methods

7.1 MODEL-SCALE EXPERIMENTS

7.1.1 General

Simulations are needed in the design and research for many purposes. Those can be utilized to study and optimize the performance of the object, study phenomena etc. In practice, it is impossible to account everything in the simulations. Thus, assumptions are needed to develop simulations or to simulate events, and these are always idealized events. However, the level of idealization varies, and the one conducting the simulation should consider what is relevant for the simulated event. As a coarse example, if one wants to simulate the bending of a cantilever beam under a mass, deadweight, it would be important to model the connection of the beam to support, the boundary condition, correctly and one would pay attention to it. However, we know that the gravity has small variation depending on the location, but for many engineering applications we would consider 9.81 m/s^2 to be an accurate value.

Simulations can be divided into numerical and experimental modelling. In some cases, the processes are too complex to be simulated numerically, or the calculation would be too expensive because of the required fidelity from the model. As an example, when the governing processes are not understood, it is very difficult or impossible to prepare proper numerical simulations. Then experimental approach gives new insight as described next.

Experimental modelling can be further divided into prototyping and scale model testing. Prototyping is a common practice in cases where the production series are long, like in car industry. However, if the products are expensive or the explored phenomenon are too dangerous to be studied firsthand, scale model testing may offer a possibility to simulate the desired event. This is schematically presented in Figure 7.1.

Ships are commonly prototypes, i.e., all built ships are final products delivered to the client because of the costs, and the series are small, especially icebreakers and passengers ships. Furthermore, phenomena related to fluid flows can be complicated to simulate. As ships operate in water, the phenomena involve simulations of a body at the boundary of two fluids – air and water. Thus, simulating ship in open-water conditions is complicated. However, simulating the breakage of ice is even more complicated that comes as an addiction to open-water simulations. Because of these reasons, model-scale testing in ice has been a widely used method in naval architecture to study the performance of a design and phenomena related to ship operations.

7.1.2 Scaling

Conducting proper model-scale tests requires correct scaling of objects and phenomena. As the tests are conducted in reduced scale, the size of the object needs to be reduced that requires attention to maintain the geometry. Furthermore, scaling laws,

DOI: 10.1201/9781003676829-7

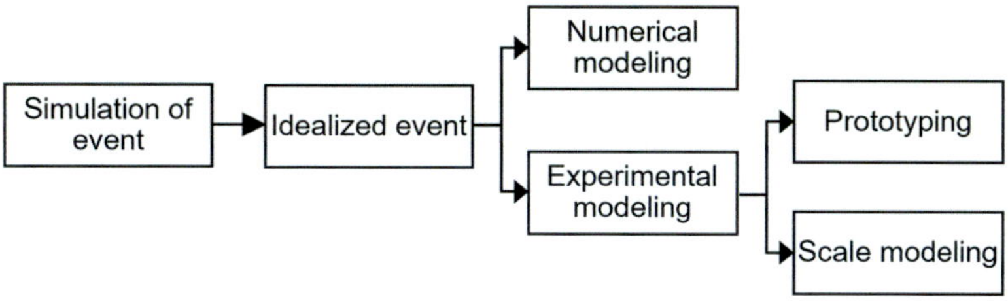

FIGURE 7.1 Schematic presentation on how model-scale testing is related to other simulations.

similarities, are needed to understand and interpret observations between scales. Generally, there are three similarities that should be maintained: geometric, kinematic and dynamic similarity. Here and later in this chapter 'model' refers to the object in reduced size while 'prototype' refers to the object that is about to be built.

Geometric similarity refers to the relationship between the dimensions. The physical model must have the same shape as the prototype, but the dimensions may be scaled by a constant factor. Kinematic similarity refers to the relationship between velocities that requires velocity at any point in the model to be proportional by a constant scale factor to the velocity at the homologous point in the prototype. Dynamic similarity refers to the relationship between forces. All forces in the model must be scaled by a constant factor to the corresponding forces in the prototype (Figure 7.2).

Scaling factor, λ, is commonly defined by the geometric scale that defines the ratio of the dimensions between the prototype, L_P, and model, L_M, $\lambda = L_P/L_M$. Dynamic similarity is defined by the relation of the most relevant forces for the process or phenomenon regarding the goals of testing. This defines the scaling law applied in the tests that is commonly expressed with the dimensionless number like Froude, Fr, or Reynolds, Re. This requires assumptions, as the applied scaling law defines the testing conditions, e.g., speed of the model, that might violate other dynamic similarities. As an example, the speed to be used in model-scale testing is different when Froude and Reynolds scaling are applied, i.e., these conditions cannot be fulfilled simultaneously. Examples of applied forces are presented below.

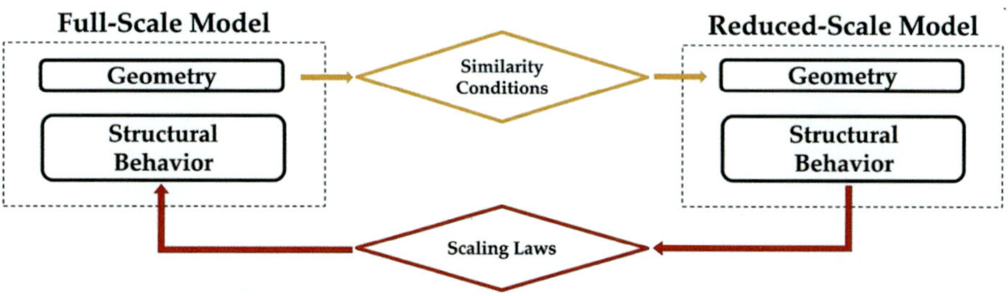

FIGURE 7.2 Relation between scales (Coutinho et al., 2016).

Inertia forces	$F_i \propto \rho L^4 T^{-2}$
Gravity forces	$F_g \propto \rho g L^3$
Viscous forces	$F_\nu \propto \nu \rho V L$
Elastic forces	$F_E \propto E L^2$
Bending forces	$F_B \propto \sigma_f L^2$
Crushing forces	$F_c \propto \sigma_c L^2$

Model-scale testing has a long history in shipbuilding. The first successful model-scale tests with ship models in open water were conducted by William Froude in 1860s, who established a methodology to predict the full-scale behaviour based on model-scale testing results. The methodology was based on maintaining Froude similitude, i.e., ensuring that the relation between inertia and gravity forces is the same in model scale and full scale. As this relation is important for modelling the ship-induced waves correctly in open-water conditions, it is still the state-of-the-art method used in model-scale testing.

Due to long traditions and solid theories in open-water conditions, model-scale testing in ice was built on the experience from open-water conditions. Several researchers have shown Froude scaling to be valid for model-scale testing in ice (Schwarz, 1983). Thus, it has been applied to scale the ship and testing condition related to kinematics and dynamics. To model the ice breaking correctly, the failure processes and forces related to the failure of ice should be represented correctly. This was achieved by maintaining the relationship between the elastic and inertia forces, represented by Cauchy similitude, between the scales (Schwarz, 1983). It is noted that the elasticity affects the characteristic length of ice that is related to the breaking pattern.

Simultaneous fulfilment of Froude and Cauchy similitude leads to the requirements that the strength and elasticity are scaled with the scaling factor, $\sigma_P/\sigma_M = E_P/E_M = \lambda$, i.e., the strength of ice needs to be reduced for model-scale testing in ice. Furthermore, the ratio between the elasticity and strength, E/σ, is considered important to model the response and failure correctly that is commonly larger than 2000 in nature. Due to non-dimensionality, the ratio should be the same in model scale. Table 7.1 presents the scaling of variables based on simultaneous fulfilment of Froude and Cauchy similitude.

Simultaneous fulfilment of Froude and Cauchy similitude is state-of-the-art method in model-scale testing in ice, and the most commonly used scaling approach. Despite different model ice types showing non-linearity in downward flexure and plasticity with small loads (von Bock und Polach et al., 2019), this scaling approach has given good results with traditional icebreaking hull forms. Partially, this has been considered to be related to the extensive experience of ice tanks in developing model ice and performing calibrations with respect to full-scale measurements (von Bock und Polach et al., 2019). However, recent research has shown that this scaling approach might not be good in model-scale tests in brash ice channel with non-conventional ice-going vessel hull forms (Matala, 2023). This research has suggested to abandon the traditional simultaneous fulfilment of Froude and Cauchy similitude, and to scale the ratio between cohesive and inertia forces of brash ice. However, this research is still ongoing and results have not been implemented extensively in practice yet.

TABLE 7.1
Scaling of Variables with Simultaneous Fulfilment of Froude and Cauchy Scaling

Variable	Scaling
Ice thickness	$[h]_P = \lambda[h]_M$
Ice comp./flex./etc. strength	$[\sigma]_P = \lambda[\sigma]_M$
Ice elastic modulus	$[E]_P = \lambda[E]_M$
Velocity	$[V]_P = \lambda^{1/2}[V]_M$
Time	$[t]_P = \lambda^{1/2}[t]_M$
Force	$[F]_P = \lambda^3[F]_M$
Mass	$[m]_P = \lambda^3[m]_M$
Acceleration	$[a]_P = [a]_M$
Frictional coefficient	$[f]_P = [f]_M$
Ice density	$[\rho]_P = [\rho]_M$

7.1.3 Model-Scale Ice

The requirement to scale down the mechanical properties of model ice used in model-scale testing in ice following Froude-Cauchy scaling approach has led to the development of different model ice. Different model ice varies in grain structure and added dopant, see Table 7.2. The different grain structure is a result of different methods in producing the model ice sheet. In practice, two methods are applied: seeding and spraying. In both cases, the basin water is first cooled down close to 0° Celsius, or slightly below zero with cold air. The air temperature is typically below −10° Celsius during the production of model ice sheet. The cooling process

TABLE 7.2
Operating Ice Tanks with the Used Model Ice. FG Refers to Fine Grained, C to Columnar, EG to Ethylene Glycol, AD to Aliphatic Detergent and S to Sugar

Facility	Structure	Chemical Composition
Aalto University, Finland	FG	Ethanol
Aker Arctic, Finland	FG	Sodium chloride
Japan Maritime United, Japan	FG	Urea
Krylov State Research Centre, Russia	C and FG	Sodium chloride
Maritime Ocean Engineering Research Institute, South Korea	C	EG/AD/S
National Research Council, Canada	C	EG/AD
National Maritime Research Institute, Japan	C	Polypropylene glycol
The Hamburg Ship Model Basin, Germany	C	Sodium chloride
Tianjin University, China	C	Urea

may produce a thin layer of ice crystals on basin water surface that needs to be cleared before the actual production of the ice sheet.

In case of seeding, after the target air and water temperatures are reached, the ice sheet production is initiated by blowing a layer of fine mist of water droplets, or solution used in the basing, over the basin. This is done after the ice crystals formed during the cooling process are cleared, or during the clearance. The fine mist lands on the surface that initiate the nucleation process and ice crystals start to form. A dense mist will lead to several simultaneous nucleation process initiations. This creates a top layer that has a small grain size and random orientation, similar to nature. The ice growth is controlled with the air temperature and duration. When cooling is continued, the best oriented crystals will continue growth downwards, blocking the ones with unfavourable orientation. This will lead to columnar grain structure that is similar to columnar ice in nature.

In case of spraying, the process begins which is similar to seeding process. However, the spraying of the aqueous solution is continued after the initial layer is seeded. When the spraying is continued, the sprayed droplets land over the already-formed layers of ice, increasing the thickness of the formed ice upwards. This leads to a model ice that has fine grain structure through the thickness. Thus, the method of producing the model ice, seeding or spraying, affects the structure of obtained model ice. Figure 7.3 gives an overall picture of the model ice in Aalto Ice tank, where spraying process is used in Aalto University to develop model ice.

FIGURE 7.3 An example of an ice basin (Aalto University) before starting the ice resistance tests with a ship model.

To be able to adjust the mechanical properties of ice, chemical dopant needs to be added in water to form chemically doped aqueous solution. Originally, salt (sodium chloride) was used as a dopant as it mimics the natural ice and enables adjusting the properties. Later, the independent development of model ice by different entities and institutions lead to usage of different chemical dopants as indicated in Table 7.2. After a dopant is added, the properties can be adjusted by altering the temperature of ice. In practice, the strength and temperature of ice are inversely related, i.e., cold ice is stronger than warm ice. Thus, the strength of ice can be adjusted to the target level by controlling the temperature of ice. However, it should be noted that different properties react to the temperature change in different phase. The way flexural strength is defined emphasizes the properties at the surface, while the compressive strength accounts for more properties through thickness. Therefore, adjusting both simultaneously correctly at the target values is not possible in practice.

As a result, the model-scale ice varies in structure and added dopant between the facilities. Different combinations lead to different ice behaviour, and certain properties are modelled better. It is commonly considered important to model as accurately as possible the ratio between flexural strength and strain modulus, as well as the ratio between compressive and flexural strength. As this ratio is dimensionless, it should be the same in natural and model ice. Lau et al. (2007) reviewed different model ice and estimated which ice represents the natural values the best. Based on the study by Lau et al. (2007), it appears that the ratio between flexural strength and strain modulus is generally modelled better by model ice having fine grain structure, while the ratio between compressive and flexural strength is better represented by columnar ice. However, it should be borne in mind that the processes and phenomena present in model-scale testing are dependent on several parameters, and the results are affected by these combination. Thus, comparing representation of single properties might lead to false conclusions. Greater emphasis should be placed on the objective of the tests, resistance or loads, and whether these can be modelled correctly.

7.1.4 Resistance in Ice

Model-scale testing in ice is commonly applied to determine the performance of a new ship concept. However, the performance can be defined in several ways, in several different ice conditions, and the tests may have significantly different scope. It could be claimed that the most common model-scale performance test in ice is to define what is the resistance of the vessel or the required power from the vessel to maintain certain speed in the defined design ice condition. This may be dictated by the ship-building contract, or maritime rules and regulations. As an example, the minimum required engine power to obtain an ice class can be determined through model-scale testing.

The performance in ice can be determined as a resistance test (ITTC, 2024) or propulsion test in ice (ITTC, 2017). Both can be used in different ice conditions, e.g., level ice, broken ice or ridged ice. Although resistance test is not directly a performance test, resistance is an essential part of ship's performance in ice, as it defines with what effective thrust a ship is able to operate, move, in ice conditions. The main purpose of a ship resistance test in ice is to define the ice resistance under given ice conditions and to evaluate the effectiveness of the hull form (ITTC, 2024).

In a resistance test, the model of the tested design is attached to a towing post and towed through the ice condition. Commonly, the attachment is made either close to the longitudinal centre of buoyancy or at the bow. The towing can be done by a rod or wire. In case a wire is used, pretension should be applied to avoid oscillations. When the model is pulled through the ice condition, the force required to pull the model through the field is measured from the towing post or from the attachment point at the model. Figure 7.4 presents an example of the test setup. The measured force equals the total resistance, R_T, consisting of open water, R_{ow}, and ice-related resistances, R_I, that are considered uncoupled and separable based on superposition principle (ITTC, 2024), see Equation (7.1). Furthermore, the ice resistance can be divided into different resistance components including breaking, R_{br}, clearing, R_c and buoyancy/static clearing, R_b, of ice (ITTC, 2024), see Equation (7.2).

$$R_T = R_{ow} + R_I \tag{7.1}$$

$$R_I = R_{br} + R_c + R_b \tag{7.2}$$

To determine the pure ice resistance from total resistance, a separate open-water resistance test needs to be performed in open-water conditions. To determine the resistance related to breaking the ice, tests can be performed in a pre-sawn ice condition. In pre-sawn ice condition, the ice field is pre-cut into a pattern that is similar to the breaking pattern of the bow. In these tests, it is assumed that in pre-sawn ice the icebreaking component is zero. Thus, the measured resistance in this condition consists of resistance related to open water, and clearing and buoyancy/static clearance of ice. Thus, the icebreaking resistance can be determined by

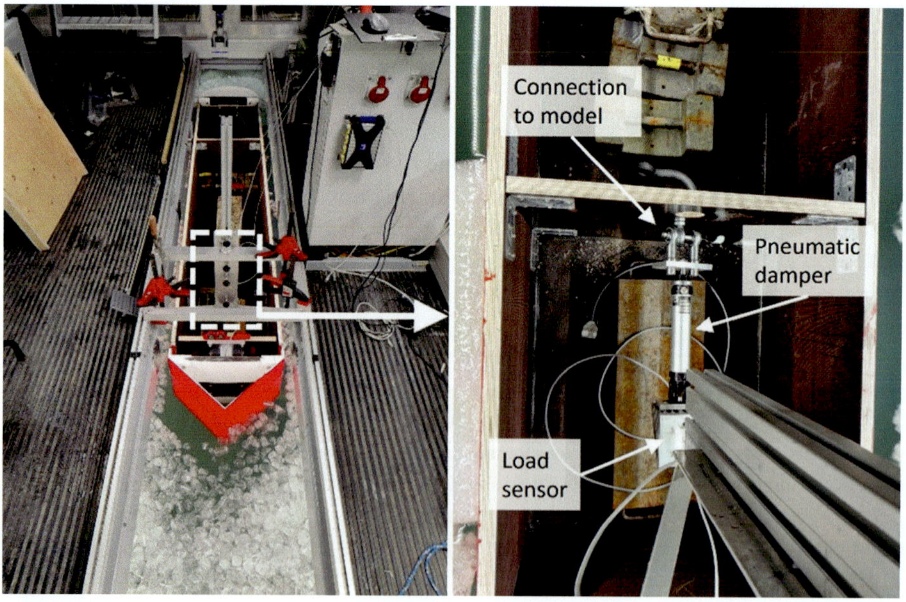

FIGURE 7.4 An example of a test setup for a towed resistance test (Suominen, 2024).

reducing the measured resistance in pre-sawn ice from the resistance measured in unbroken level ice, see Equations (7.1) and (7.2).

Commonly, the operational speeds in ice conditions are small compared to the speeds in open water. As the resistance in open water increases exponentially as a function of speed, the open-water resistance is commonly very small compared to ice resistance at the speeds used in model-scale testing in ice. The significance of the icebreaking resistance depends on the hull form. In case the ship operates in an intact ice condition, the breaking component is never insignificant, but its relative importance can be reduced by good icebreaking designs. Figure 7.5 shows measured towing force in open-water condition, pre-sawn ice and level ice.

Modelling the ice conditions and properties accurately is difficult and small variations in the model ice field will exist in any case. Thus, the measurements need to be corrected to the target ice conditions. Commonly, ice thickness, strength and friction coefficient needs to be corrected to the target values. Thus, measuring these from the ice field spatially and temporarily close to the tests is important. When these are measured, the measurements can be corrected with the established methods as described in ITTC (2024), or as defined by the testing facility. In practice, if deviations from the target ice thickness occurs, the measured ice resistance is corrected by multiplying the resistance by the ratio of target and measured ice thickness, h_{target} and h_{meas} respectively, raised to the exponent x. The exponent x typically varies between 1.0 and 2.0 and is defined by the test basin (ITTC, 2024). Deviations in strength is corrected by multiplying the resistance component related to breaking by the ratio of target strength and measured strength, σ_{target} and σ_{meas}, respectively. The deviation in friction coefficient is corrected by multiplying the ice resistance with a correction factor, C_μ, defined by the ice basin. Thus, the corrected resistance, $R_{I,\text{cor}}$, can be expressed as:

$$R_{I,\text{cor}} = C_\mu \left(\left(\frac{\sigma_{\text{target}}}{\sigma_{\text{meas}}} \right) R_{br} + R_r \right) \left(\frac{h_{\text{target}}}{h_{\text{meas}}} \right)^x \tag{7.3}$$

where R_r includes resistance related to clearing and submersion, i.e., $R_r = R_c + R_b$.

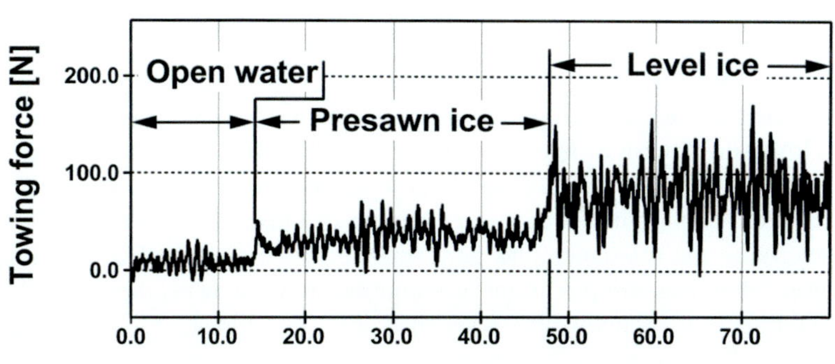

FIGURE 7.5 An example of resistance measurements in open water, pre-sawn ice and level ice (ITTC, 2002).

7.1.5 Propulsion Test in Ice

As mentioned in the previous chapter, ship's performance in ice can be determined through propulsion test in ice. The main difference to the resistance test in ice is that the model is equipped with a propulsion system that generates thrust during the test. The purposes of performing propulsion tests are generally: (i) to determine the attainable speed, resistance and required propulsion power in specific ice conditions, (ii) to determine the effectiveness of the combined hull-propulsion system in ice and (iii) to investigate the ice interaction with appendages and propulsion system (ITTC, 2017).

The propulsion test can be done as a towed propulsion test in ice or open water, or as a free running model test. In case of a towed propulsion test, the test setup is similar to resistance test, i.e., the model is attached to the carriage and moved through the ice condition with a constant speed. As a difference, the model is equipped with a propulsion system that generates a thrust, T, that pushes the model forward with an effective thrust, Teff, that accounts for the influence of the hull on the thrust through thrust deduction, t,

$$T_{\text{eff}} = T(1 - t) \tag{7.4}$$

As the generated effective thrust assists the model in proceeding through the test conditions, the measured force, Fx, depends on the open-water resistance, ice resistance, and the effective thrust.

$$Fx = R_{ow} + R_I - T_{\text{eff}} \tag{7.5}$$

In case the effective thrust is smaller than the total resistance, the carriage pulls the model through the condition. However, if the effective thrust is larger than the total resistance, the model pushes the carriage. Thus, the model needs to be attached to the carriage in a manner that allows pushing and pulling of the model. The tests are commonly done with different thrusts using different *rps* and torque. This allows the evaluation of the self-propulsion point, i.e., where Fx equals zero. When the self-propulsion point is known, it is possible to determine the required propeller shaft power, $P_{D,\text{req}}$, from

$$P_{D,\text{req}} = 2\pi Q_{\text{req}} n_{\text{req}} \tag{7.6}$$

It should be noted that the propulsion tests in ice can simulate a realistic view of interaction between ice and propulsors (such as ice milling, blockage and proximity effect). However, in case of heavy propeller-ice interaction, direct use of ice torque which is mainly from ice milling may lead to inaccurate value due to inappropriate scaling of propulsion unit and crushing strength of model ice (ITTC, 2017).

To determine the resistance and performance in ice through free running tests, towed propulsion tests in open water are required. In these tests, the available thrust after overcoming the open-water resistance, i.e., net thrust, T_{net}, is defined for different propeller *rps* with different model speeds. As a result, a set of curves for different *rps* are gained, where the available thrust is presented as a function of speed. When a free-running test is conducted under ice conditions, the *rps* level is set and the speed that the model is able to achieve in the testing conditions is recorded. When the *rps* is known and the speed is determined through the test, it is possible to check the net thrust for this combination of *rps* and speed from the open-water curves. The defined net thrust equals the ice resistance in the testing condition with the obtained speed. Here, it is assumed that the propeller performance is not affected by ice.

7.1.6 Local Loads on Ship's Hull

Local loads on ship hull can also be measured in model-scale tests in ice. Here, the local ice loads on a ship hull are defined as the loads occurring on a frame or web-frame area in full scale. However, these measurements are relatively rare and commonly conducted as a part of performance tests. Thus, the scaling and testing conditions are commonly the same as in performance testing. As the ice properties are scaled down, the local pressures on the hull are significantly reduced, making the conditions easier for the applied measuring devices.

The local ice loads on the ship hull have been measured at model scale using tactile sensor sheets that are installed on the hull (see, e.g., Kujala and Arughadhos, 2012; Suominen and Kujala 2012, 2013; Izumiyama et al., 2007), or using traditional load or force sensors through cut-outs on the model hull (see, e.g., Suominen and Kujala 2012, 2013; Myland and Hisette, 2021). The benefit of using tactile sensor sheets is the possibility to measure loads with high spatial accuracy. The sensors can measure loads from an area of 100 cm^2 with a spatial accuracy of 1 mm^2. Note that these are order of magnitude and not exact numbers. In addition, as the loads are measured on the hull, the gained pressure is from the interface of the structure-ice interaction. Thus, the measured loads represent the forces acting directly on the hull. A drawback of these devices is that their temporal resolution (i.e., sampling frequency) and measurement accuracy have been criticized in the past. Thus, if these are used in the tests, a proper calibration and verification of the accuracy and spatial resolution is important.

The benefits and drawbacks of using traditional load or force sensors through cut-outs are close to opposite to tactile sensor sheets. In general, the cut-outs are always relatively large compared to the possibilities with tactile sensor sheets. The size of the cut-outs is measured in centimetres, meaning the representative extent in full scale is usually web-frame spacing or larger area. Thus, the measured loads are often semi-local. Additionally, as there is a structure between the sensor and ice loads, the structure may affect the result, e.g., by damping the loading. The benefit of using load or force sensors is that the temporal resolution and the accuracy are often at very high level.

7.2 FULL-SCALE TESTS OF ICE-INDUCED LOADS ON SHIP HULL

Full-scale ice-induced loads can be measured by using direct or indirect measurement methods. Here, the direct measurement methods refer to utilizing external devices that are mounted or integrated in the structure to measure loads. The indirect methods utilize inverse engineering where the external loading is defined based on measured structural response like strain or displacement.

As ice-induced load measurements on ship hull have long traditions, different approaches have been applied to measure the loads. Vuorio et al. (1979) developed a pressure gauge system that defines the ice pressure by measuring the pressure from internal surface of the hull within the area of a stiff structure mounted inside the hull. Glen and Blount (1984) employed through hull-mounted pressure gauges that were measuring the pressure on the gauge area. Hoffmann (1985) utilized load sensors in a load panel that was specially built for the hull. The measurements obtained by these systems could be considered accurate. The drawback of these systems is that the measurements are very local, and the systems developed by Vuorio et al. (1979) would require extensive installation work. The load panel utilizing load sensors in a specially build section in the outer hull (Hoffmann, 1985) would require even more extensive work and structural alteration. Furthermore, no result has been published from the load panel indicating that it has not been successful.

Understanding the external pressure induced by ice, Riska et al. (1990) installed a PVDF film and a window on the outer hull to measure local pressures and observe the local processes. As the pressure in the interaction can be extremely high and mechanical wear is significant, the external gauge did not last long, but the window enabled qualitative observations. Gagnon et al. (Gagnon, 2008, Gagnon et al., 2020) have developed optics-based methods to measure the ice-induced pressure that can withstand the pressure. This method utilizes acrylic beams and is based on defining the deformation of these beams under loading, i.e., it could be understood as indirect method. This method has been shown to be able to provide a high-resolution pressure distribution from the contact. The drawback of the system is that it requires the system to be installed on the ship that requires structural alteration. Furthermore, as it alters the structural stiffness, it is not sure how well the measured pressures can be generalized to the actual structure of the ship. Pictures from the direct load measurement installations are presented in Figure 7.6.

Despite direct methods being applied, indirect methods have been more common in determining the local ice-induced loads. These methods determine the load affecting the structure by measuring the structural response. Traditionally, the structural response has been measured with strain gauges although applicability of other methods has been studied. The most common approach is to measure the response of frames or stiffeners. In these cases, the external loading is determined from the load-strain relation, where the structural response is determined based on an assumed load location, shape and pressure distribution (see, e.g., Riska et al., 1983; Ralph et al., 2003). These are commonly kept constant through the measurements. Attempts to determine the local loading by measuring the bending strain at the flange of the frame have been made, but these measurements have been difficult to interpret.

On a few occasions, these measurements have been based on the compressive strain on the frame normal to the shell: USCGC Polar Sea (St. John et al., 1984; Daley et al., 1990), RV Nathanial B. Palmer (St. John and Minnick, 1995). Here, the aim is to measure the localized load affecting the structure, as the strain normal to the

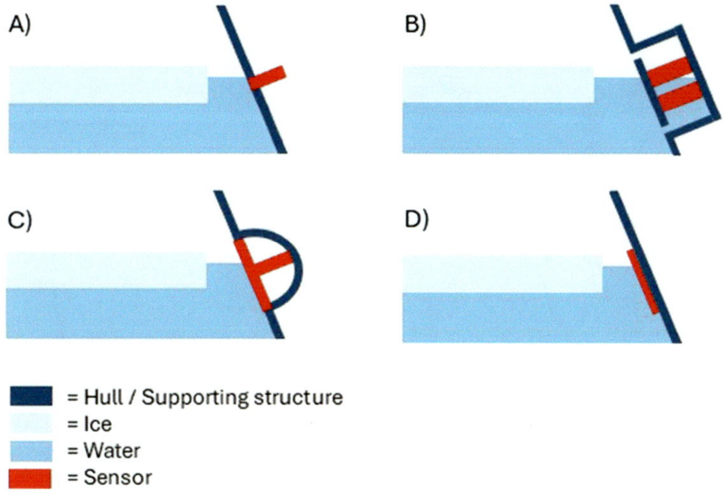

FIGURE 7.6 Principle ideas of direct measurement methods by (a) pressure gauges, e.g., by Glen and Blount (1984), (b) separate load measuring device, e.g., by Hoffmann (1985), (c) ice pressure gauges, e.g., by Vuorio et al. (1979) and (d) PVDF film, e.g., by Riska et al. (1990).

hull plating decreases rapidly outside the contact area. Thus, with this approach, the loads that have a direct effect at the location of the strain gauge can be determined. More commonly, these are based on the measured shear strain difference between two locations on a frame, see e.g. Varsta (1976), Kujala and Vuorio (1986), Ghoneim and Keinonen (1983), St. John et al. (1994), Ritch (2008), Leira et al. (2009), Suominen et al. (2013), see also Figure 7.7 as an example of the typical instrumentation. The idea in this approach is that the ice-induced load acts as a shear force on the frame. Thus, the ice-induced load between two locations, upper and lower frame parts, can be determined by measuring the shear strain difference between these locations.

Figures 7.8 and 7.9 gives more insight to the basic principle of shear strain gauges. When we measure the shear strain on both side of the loading, we can define the shear force, i.e., ice load, acting between the measurement points as illustrated in Figure 7.8. Figure 7.9 presents unloaded and loaded stiffened plating. The principle of measuring the shear strain difference is the same as in the case of single frame. However, as indicated in Figure 7.9, the load is distributed to the adjacent frame. This load transportation between frames through plating needs to be accounted for that can be done utilizing the inverse coefficient matrix as described next.

It is well known that the magnitude, location and load patch dimensions of ice-induced loading vary rapidly in full-scale measurements, see e.g. Kujala (1994). The measurement uncertainty in these approaches is related to the assumed loading conditions. From these two approaches, the difference in the shear strain is less sensitive to the variation in the loading condition than the measurements based on the compressive strain normal to the hull, as it is less sensitive to the change in load

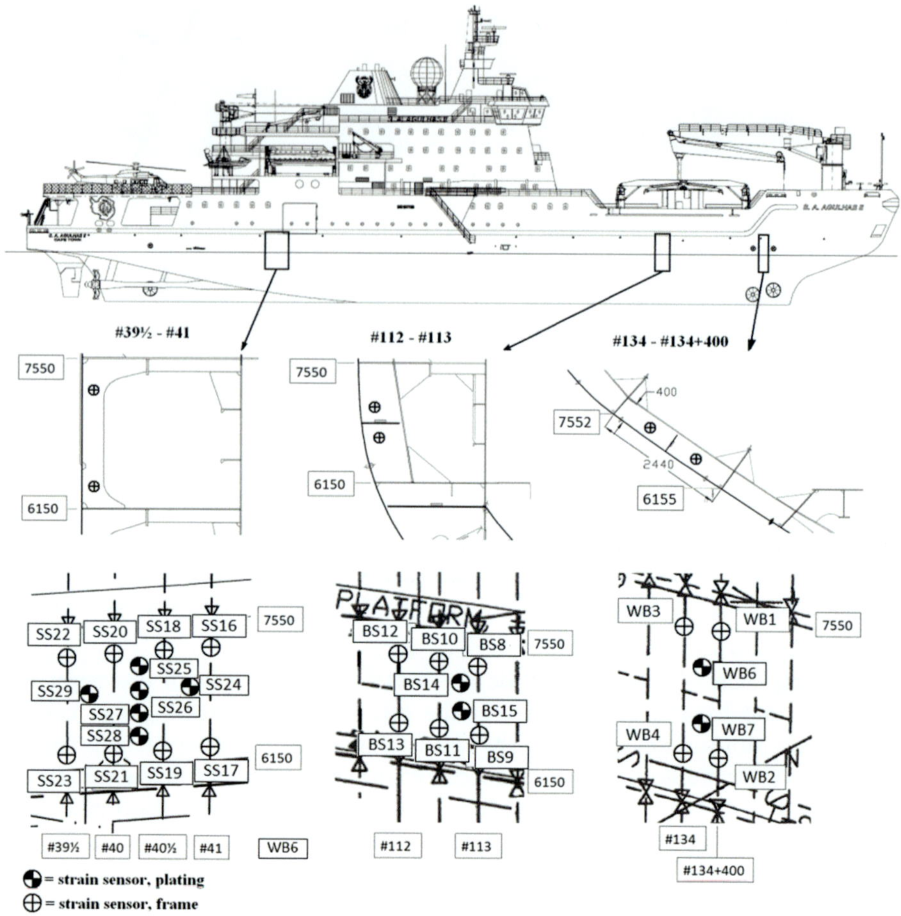

FIGURE 7.7 An example of a typical full-scale instrumentation onboard *SA Agulhas II.*

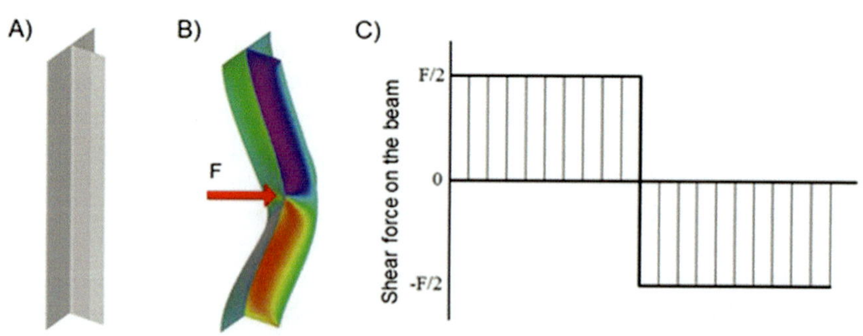

FIGURE 7.8 (a) Unloaded frame; (b) a frame loaded with point load F at the mid-span, the colour indicates shear strain on the frame and (c) idealized shear force distribution on the frame.

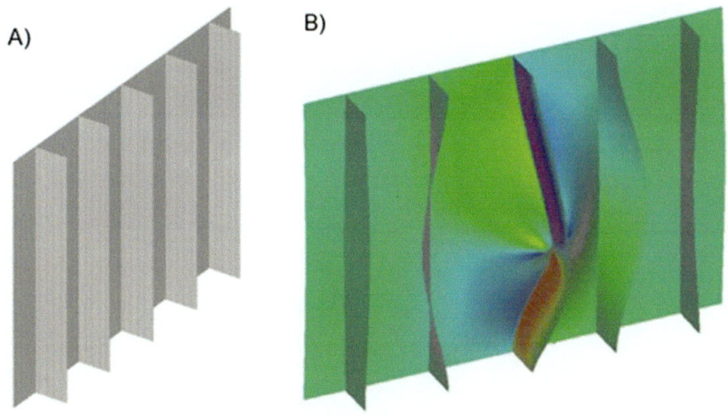

FIGURE 7.9 (a) Unloaded stiffened plating; (b) a stiffened plating load with a point load, the colour indicates shear strain on the frame.

patch height and location on the frame assuming the loading is affecting between the gauges (Suominen, 2018). However, the shear strain difference-based method is sensitive to the horizontal dimensions of the loading that depends on the structural arrangement that can be reduced by extending the instrumentation to the adjacent frames (Suominen et al., 2017). In these cases, the response of the neighbouring structure can be accounted for by constructing a response matrix that represents the contribution of adjacent frames to the load determination on a frame.

The response of the adjacent frames can be accounted for in the full-scale measurements by introducing an influence coefficient matrix. The external load on the structure is determined from the measured shear strain difference with an influence coefficient matrix a and its inverse, c, using the relations

$$\{F\} = [a]\{\Delta\gamma\} \text{ and } [c_{ij}] = \Delta\gamma_i(F_j) \tag{7.7}$$

where $\Delta\gamma_i$ $(i = 1 \ldots m)$ is the shear strain difference measured on the frame i, m is the number of instrumented frames, and F_j is the external force exerted on the frame j. The diagonal terms define the force-strain relation of the frame under loading and the off-diagonal terms determine the response of the adjacent frames. Thus, the diagonal terms are only affected by the force carried by the loaded frame, while the off-diagonal terms consider the shear force and torsion of the adjacent frames. This can be shown with an analytical description of the matrix (see Chapter 2.2. in Suominen [2018]).

The influence coefficient matrix a is the inverse of the matrix c. It is a common practice to determine the influence coefficients with Finite Element Analysis, FEA (Ralph et al. [2003]; Suominen [2018]), as it is considered more accurate. However, the inverse methods have their own uncertainties and methodological problems and are under development. For these reasons, uncertainty in the measurements should be studied.

In addition to using finite element modelling to calibrate the measuring system, physical calibration can also be used, as shown in Figure 7.10.

FIGURE 7.10 Physical calibration of the measuring system onboard *SA Agulhas II*.

7.3 ICE-LOAD EVALUATION USING DAMAGE STATISTICS

7.3.1 Accident and Ship Hull Damage Statistics

There is a long history of Arctic marine transport. For example, year-round navigation has been maintained since 1978–1979 in the ice-covered western regions of the Northern Sea Route (between the port of Dudinka on the Yenisei River and Murmansk), see e.g. AMSA (2009). The Arctic today is destinational, conducted for community re-supply, marine tourism and moving natural resources out of the Arctic. Regions of high concentrations of Arctic marine activity occur along the coasts of northwest Russia, and in the ice-free waters off Norway, Greenland, Iceland and in the U.S. Arctic. Significant increases in cruise ships, a majority not purpose-built for Arctic waters, have been observed in the summer season around Greenland within the past decade.

Comprehensive statistics on accidents in the Arctic waters are presently not easy to find. However, there have been several attempts to compile and analyse statistics on Arctic accidents. Each attempt takes a slightly different approach than the other, making it more difficult to draw general conclusions from them all. For instance, each study defines the Arctic geographic area in a slightly different manner than the other. Some studies focus strictly on ship damage, while others include personal injuries, and each takes a different approach to categorizing and explaining Arctic accidents.

Kubat and Timco (2003) compiled and analysed a database of ship damage in the Canadian Arctic for the period between 1978 and 2002. They focused on Canadian waters north of 60°N latitude. They analysed 125 events that occurred during that time frame and concluded that the presence of multiyear ice was a factor in 73% of the damaging events, with first-year ice being responsible for the remaining damaging events. Also, the damage was more severe when multiyear ice was present.

Marchenko's (2012) book on the Russian Arctic waters takes a different approach. She goes into some detail regarding accidents in the Arctic seas bordering Russia. Her coverage spans accidents since 1900 up until the early 1990s, when the volume of Arctic traffic in Russian waters dropped off dramatically due to the collapse of the Soviet Union. However, most of the incidents described (some in great detail) occurred before the 1980s, and while they are certainly informative, it is unclear what lessons might be learned from some of these incidents given that many of them predate the use of steel ships and diesel/electric/nuclear-powered propulsion. Of the handful of the more recent events described in this book, a common theme is the difficulty of dealing with ice and the problems that navigation in ice presents – both with and without icebreaker assistance.

As part of the Arctic Council's Arctic Marine Shipping Assessment (AMSA) 2009 Report, a database of a summary of the incidents and accidents occurring in the Arctic region between 1995 and 2004 was developed from several other existing datasets. While this dataset is one of the more comprehensive available, it does not include any incidents that occurred in Russian waters. A lack of recent data from Russia among any of the examined sources is regrettable. The AMSA covers incidents in what is called the 'circumpolar region', which is not further defined – leaving a question as to how loosely defined is AMSA's definition of the Arctic geographic region. A summary of results from AMSA can be found in Table 7.3, which breaks down the accidents by vessel type and cause of damage, respectively.

While this summary is a good starting point, we believe that further analysis of the AMSA accident database might provide some useful insight into the nature of Arctic accidents and how they might be prevented. For instance, two prior studies cited the presence of ice as a major contributing factor in Arctic accidents, but this is not a focus of the AMSA. AMSA does acknowledge that further work is required and it cites the development of a consistent and accurate circumpolar database of Arctic ship activity along with ship accidents and incidents to date as an area where further research would be useful. Goerlandt et al. (2017) have taken a step in this direction by analysing wintertime accidents in the Northern Baltic based on AIS data together with ice and weather information data from SMHI, aiming to provide qualitative insights in patterns and outlier cases in the accidental conditions. It would probably be difficult to get the same level of detail for a similar analysis for the Arctic, but it is a possible path for future research.

The challenge with all these damage databases for the Arctic waters is that they are not detailed enough to evaluate the sequence of events causing the accidents, and therefore they cannot be easily applied in the development of risk-based design methodologies. In addition, the configuration of the damage is typically not reported to the extent needed for proper analysis of the load levels required to cause the damage. This is an important gap as it means there is still considerable uncertainty on the estimated maximum ice loads on the ship hull.

Qualitative assessments are well suited for obtaining an overall understanding of the risks associated with a specific type of operation, or when sufficient data are not

TABLE 7.3

Number of Accident as a Function of Ship Types and Primary Reason of the Accident (AMSA, 2009)

Vessel Type	Number of Accidents	Percentage
(a)		
Bulk Carrier	37	12.6
Container Ship	8	2.7
Fishing Vessel	108	36.9
General Cargo Ship	72	24.6
Government Vessel	11	3.8
Oil/Gas Service & Supply	1	0.3
Passenger Ship	27	9.2
Pleasure Craft	0	0.0
Tanker Ship	12	4.1
Tug/Barge	15	5.1
Unknown	2	0.7
Total	**293**	**100**
Primary Reason	**Number of Accidents**	**Percentage**
(b)		
Collision	22	7.5
Damage to Vessel	54	18.4
Fire/Explosion	25	8.5
Grounded	68	23.2
Machinery Damage/Failure	71	24.2
Sunk/Submerged	43	14.7
Miscellaneous	10	3.4
Total	**293**	**100**

available to support a quantitative assessment. To assess the risks related to Arctic maritime operations, we believe that the use of historical data can help designers to identify typical operational and specific accidental events, and to roughly assess their likelihood. Therefore, Huuhtanen (2018) compiled all publicly available Arctic maritime accident data that could be identified into a comprehensive database. In total, we found data on some 3362 accidental events from the period 1975–2018 involving ships above 500 GT. An overview of the geographical distribution of reported events is presented in Figure 7.11.

Data analysis resulted in the following findings, among others (Huuhtanen, 2018):

- Heavy weather is behind most total ship losses, and behind most casualties.
- Occupational (onboard) accidents are the most common type of accident resulting in at least one fatality.

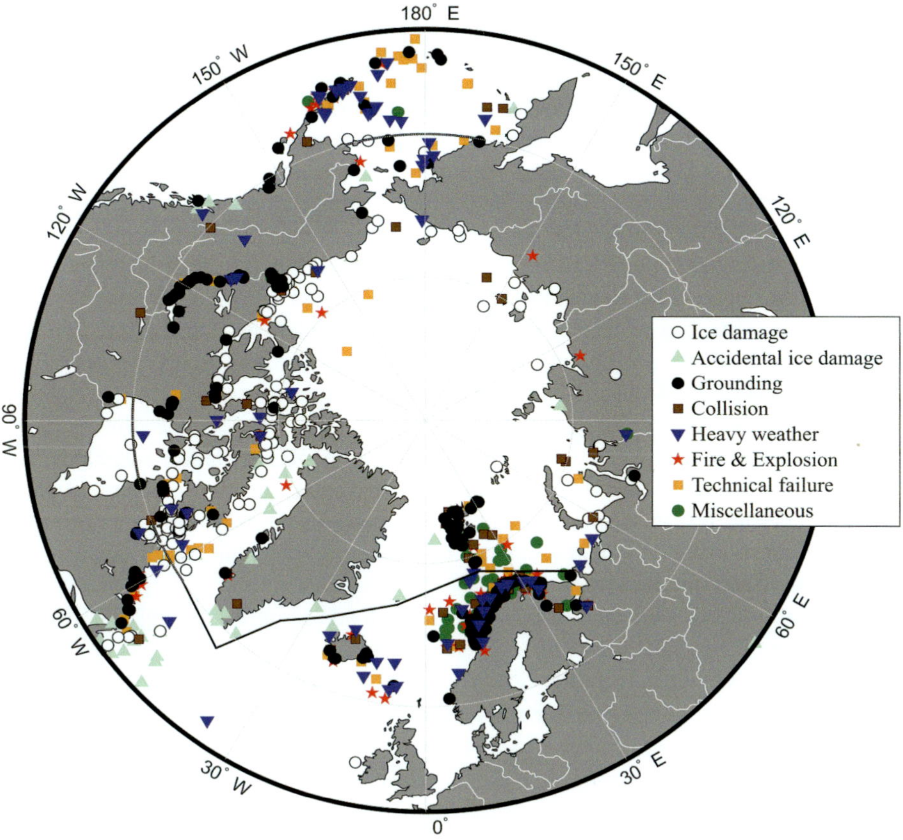

FIGURE 7.11 Calculated ice loads for different hull areas, return periods, and operating profiles (Huuhtanen, 2018).

- The frequency of occurrence of different types of accidents depends on the sea area (e.g., we find accidental ship-ice interactions are common in the Canadian Archipelago, that groundings are common in the Svalbard area, and that heavy weather accidents are common in the Norwegian Sea).
- Most ship collisions in the Arctic occur in ice-free waters.

7.3.2 Evaluation of the Ice-Induced Loads from Damages

In the development of the Finnish-Swedish ice-class rules for the Baltic Sea, damage statistics due to the ship-ice interaction have played an important role, see Section 6.2.1. There are extensive databases of detailed damage cases starting from the early work of Johansson (1967) and continuing work by Kujala (1991) and Hänninen (2004). They cover ships only navigating in the Baltic Sea, meaning operations in first-year ice with a maximum thickness of around 1.0 m. Hindley and Ranki (2009) made an extensive analysis of ice-induced damages on the vessels navigating in Arctic waters, the analysis include vessels like *Lunni*, *Uikku*, *SA-15*, *Norilsk Nickel* and *Antarcticaborg*. They concluded a potential reduction in the 'stern as a bow' ice-strengthening requirements for the stern region of the ship designed to operate stern first in ice and an increase in the ice strengthening for aft shoulder for ships designed to operate stern first in ice.

In the following section, a detailed analysis is shown on how the damage statistics can be used to evaluate the load level caused by the damage, based on the work by Kujala (1991) and Valkonen (2006). Figure 6.3 shows a typical ice-induced damage occurring in the ice conditions of the Baltic Sea. 'Damage' is perhaps too strong a term in this context, because it is more about a small dent in the side of the ship. These are very rarely safety risks and they are usually only fixed during normal docking.

When the shape of the dent is known, it is possible to evaluate how big a force has acted on the structure to cause the damage. The problem is that the location and shape of the load are not known, but by assuming, for example, the height of the load, the pressure load and the width of the pressure load that cause the damage can be found. Since dents in the side of the ship are dealt with, plastic dimensioning methods must be used in the load assessment. The plastic failure mechanism is different for different structural parts: longitudinally and transversely framed plate, longitudinal and transverse frames and web-frames. In the following section, only the calculation of plate damage is discussed, since the vast majority of the detected damage has been in the plating.

Hayward's (2001) method of estimating the loads required to cause permanent deformation of the plate is the most recent approach in this area. He has conducted mostly finite element calculations by using the following formula to estimate the load, when the permanent deflection w_p of the plate is known.

When $w_p/t < \; = 1.0$:

$$q = \frac{p_c h_C}{f_{DL}} \left[1 + \frac{w_p^2}{3t^2} \left(\frac{\zeta_0 - (3 - 2\zeta_0)^2}{3 - \zeta_0} \right) \right] \tag{7.8}$$

When $w_p/t > 1.0$:

$$q = \frac{2 p_c h_C w_p}{t f_{DL}} \left[1 + \frac{\zeta_0 (2 - \zeta_0)}{3 - \zeta_0} \left(\frac{t^2}{3 w_p^2} - 1 \right) \right] \tag{7.9}$$

where h_c is load height and p_c is a constant pressure causing the collapse of the plating following the so-called letter envelope (Jones, 1976):

$$p_c = \frac{48 M_P}{s^2 \left(\sqrt{3 + \left(\frac{s}{l}\right)^2} - \frac{s}{l} \right)^2} \tag{7.10}$$

t is plate thickness and f_{DL} is obtained for the longitudinally framed plating from:

$$f_{DL} = -0.6263 {x_L}^2 + 1.5363 x_L \tag{7.11}$$

and x_L is:

$$x_L = \frac{h_C}{s} \left(\frac{s}{t} \frac{s}{l} \right)^{0.1} \tag{7.12}$$

s is frame spacing and l is the span of the longitudinal frame.

In addition, the parameter ζ_0 is obtained from:

$$\zeta_0 = \frac{s}{l}\left(\sqrt{3 + \left(\frac{s}{l}\right)^2} - \frac{s}{l}\right) \tag{7.13}$$

and M_p:

$$M_P = \sigma_y \frac{t^2}{4} \tag{7.14}$$

where σ_y is the yield strength of the used material.

For transversely framed structures, the formulations can be updated just by changing the parameter f_{DL} to f_{DT}:

$$f_{DT} = -0.1330{x_T}^2 + 0.6701x_T \tag{7.15}$$

where

$$x_T = \frac{h_C}{s}\left(\frac{s}{t}\right)^{0.2} \tag{7.16}$$

Table 7.4 shows a summary of the statistical ice damage for transversally and longitudinally framed ships. Table 7.4 and formulas 7.1–7.16 give an estimate of the required ice load as shown in Figure 7.12. Especially with long load lengths ($L > 2$ m), the load values calculated based on ice damage are surprisingly large. Most of the ice damage has occurred in the midship or stern as a result of loads caused by compressive ice. This indicates that in a compression situation the loads can be clearly greater than, for example, in the bow of an icebreaker when it is breaking ice. However, it must be remembered that there is often a lot of inaccuracy in measuring ice damage, which also results in inaccuracy in the values shown in Figure 7.12.

TABLE 7.4

A Summary of the Ice-Induced Damages on the Transversely and Longitudinally Framed Structures (Kujala, 1991)

Ship Number	Frame Spacing [mm]	Frame Span [mm]	Plate Thickness [mm]	Maximum Deflection [mm]	Yield Strength [MPa]	Design Pressure [MPa]
(a)						
4	400	3000	19	20	290	1.74
7	380	3000	16.5	10	290	1.48
13	350	3200	16	20	290	1.50
	700	3200	10	15	290	1.11
14	350	2375	14	15	290	1.33
	700	2375	13	25	290	1.31
16	350	3500	10	25	290	1.22

(Continued)

TABLE 7.4 (Continued)

Ship Number	Frame Spacing [mm]	Frame Span [mm]	Plate Thickness [mm]	Maximum Deflection [mm]	Yield Strength [MPa]	Design Pressure [MPa]
21	400	2200	19.5	10	290	1.78
	800	2200	19.5	15	290	1.74
	400	2200	13.5	10	290	1.78
	800	2200	13.5	15	290	1.74
35	350	2500	15.5	15	290	1.18
43	400	1750	17	25	350	1.49
44	400	1750	9.5	30	350	1.49
51	350	2100	15.5	15	290	1.46

Ship Number	Frame Spacing [mm]	Frame Span [mm]	Plate Thickness [mm]	Maximum Deflection [mm]	Design Pressure [MPa]
(b)					
7	450	3040	20	10	1.43
13	350	2800	10	20	1.48
45	300	2800	14.5	24	1.57
	300	2800	14.5	30	
48–50	343	2100	17	30	1.76

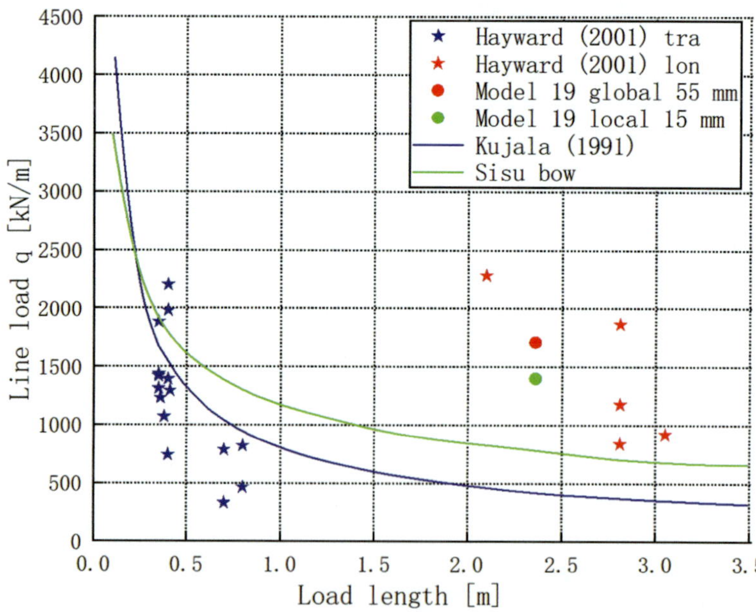

FIGURE 7.12 Ice load calculated based on ice damage statistics as a function of load length (re-produced from Valkonen, 2006). 290 MPa has been used as the yield strength of steel. The picture also shows the maximum ice load measured in *JM Sisu*'s bow and the maximum load curve presented by Kujala (1991) based on an earlier analysis of ice damage.

7.4 SUMMARY

Model-scale testing in ice has been a widely used method in naval architecture to study the performance of a design and phenomena related to ship operations in ice. This chapter summarises the basic principles of scaling, various methods to make model ice and the test principles for ship resistance in ice and for ship propulsion tests as well how to measure local pressures on the model ship hull.

Full-scale ice-induced loads can be measured by using direct or indirect measurement methods. Here, the direct measurement methods refer to methods utilizing external devices that are mounted or integrated in the structure to measure loads. The indirect methods utilize inverse engineering where the external loading is defined based on measured structural response like strain or displacement. It is well known that the magnitude, location and load patch dimensions of ice-induced loading vary rapidly in full-scale measurements. The measurement uncertainty in these approaches is related to the assumed loading conditions. The difference in the shear strain is less sensitive to the variation in the loading condition than the measurements based on the compressive strain normal to the hull, as it is less sensitive to the change in load patch height and location on the frame assuming the loading is affecting between the gauges. However, the shear strain difference-based method is sensitive to the horizontal dimensions of the loading that depends on the structural arrangement that can be reduced by extending the instrumentation to the adjacent frames. In these cases, the response of the neighbouring structure can be considered by constructing a response matrix that accounts for the response of adjacent frames to the load determination on a frame.

Extensive accident databases exist. The challenge with all these damage databases for the Arctic waters is that they are not detailed enough to evaluate the sequence of events causing the accidents, and therefore they cannot be easily applied in the development of risk-based design methodologies. In addition, the configuration of the damage is typically not reported to the extent needed for proper analysis of the load levels required to cause the damage. This is an important gap as it means that there is still considerable uncertainty on the estimated maximum ice loads on the ship hull. When the shape of the dent caused by ice is known, it is possible to evaluate how big a force has acted on the structure to cause the damage.

REFERENCES

Arctic Council, 2009. Arctic Marine Shipping Assessment (AMSA) 2009 Report.

Coutinho, C., Baptista, A., and Dias Rodrigues, J., 2016. Reduced scale models based on similitude theory: A review up to 2015. *Engineering Structures*, 119, 81–94.

Daley, C., St. John, J., Brown, R., and Glen, I., 1990. *Ice loads and ship response to ice – Consolidation report*. Ship Structural Committee, Report No 340.

Gagnon, R., 2008. Analysis of data from bergy bit impacts using a novel hull-mounted external Impact Panel. *Cold Regions Science and Technology*, 52, 50–66.

Gagnon, R., Andrade, S., Quinton, B., Daley, C., and Colbourne, B., 2020. Pressure distribution data from large double-pendulum ice impact tests, *Cold Regions Science and Technology*, 175, 103033.

Ghoneim, G., and Keinonen, A., 1983. Full-scale impact tests of CANMAR Kigoriak in thick ice. In *Proceedings of the 7th Int. Conf. on Port and Ocean Engineering under Arctic Conditions, POAC*, Espoo, Finland, Vol. 3, March 21–31, 1983, pp. 329–346.

Glen, I., and Blount, H., 1984. Measurement of ice impact pressures and loads onboard CCGS Louis S. St. Laurent. In *3rd Conference on Offshore Mechanics and Arctic Engineering*, New Orleans, pp. 246–252.

Goerlandt, F., Montewka, J., Zhang, W., and Kujala, P., 2017. An analysis of ship escort and convoy operations in ice conditions. *Safety Science*, 95, 198–209.

Hänninen, S., 2004. Incidents and Accidents in Winter Navigation in the Baltic Sea, Winter 2002–2003 (unpublished). Winter Navigation Research Board. Research Report No 54.

Hayward, R., 2001. *Plastic response of ships shell plating subjected to loads of finite height.* Master's Thesis, Memorial University of Newfoundland, St. John's, New Foundland.

Hindley, R., and Ranki E., 2009. Learning from ice damage investigations for future Arctic ships. In *ICSOT 2009: Ice Class Vessels*, Busan, Korea.

Hoffmann, L., 1985. Impact forces and friction coefficients on the forebody of the German polar research vessel POLARSTERN. In *Proceedings of the 8th Int. Conf. on Port and Ocean Engineering under Arctic Conditions, POAC*, Narssarssuaq, Greenland, Vol. 3, pp. 1189–1202.

Huuhtanen, J., 2018. *Collection and analysis of arctic maritime accident data.* Espoo: Aalto University.

ITTC, 2002. *ITTC - Recommended procedures and guidelines.* Procedure, Resistance Tests in ice. 7.5-02-04-02.1.

ITTC, 2017. *ITTC - Recommended procedures and guidelines.* Procedure, Propulsion Tests in ice. 7.5-02-04-02.2.

ITTC, 2024. *ITTC - Recommended procedures and guidelines.* Procedure, Resistance Tests in ice. 7.5-02-04-02.1.

Izumiyama, K., Takimoto, T., and Uto, S., 2007. Length of ice load patch on a ship bow in level ice. In *10th International Symposium on Practical Design of Ships and other Floating Structures (PRADS)*, Houston, TX.

Johansson, B., 1967. On the ice-strengthening of ship hulls. *International Shipbuilding Progress*, 14, 231–245.

Jones, N., 1976, Plastic behaviour of ship structures. *SNAME Transactions*, 84, 115–145.

Kubat, I., and Timco, G., 2003. Vessel damage in the Canadian arctic. In *Proceeding 17th International Conference Port and Ocean Engineering under Arctic Conditions, POAC'03*, pp. 203–212.

Kujala, P., 1991. Damage statistics of ice-strengthened ships in the Baltic Sea 1984–1987. Winter Navigation Research Board. Report No. 50. 61 p. + app. 5 p.

Kujala P., 1994. *On the statistics of ice loads on ship hull in the Baltic.* Doctoral Thesis. Helsinki University of Technology, Ship Laboratory, 98 p.

Kujala, P., and Arughadhoss, S., 2012. Statistical analysis of ice crushing pressures on a ship's hull during hull-ice interaction. *Cold Regions Science and Technology*, 70, 1–11.

Kujala, P., and Vuorio, J., 1986. *Results and statistical analysis of ice load measurements on board icebreaker Sisu in winters 1979 to 1985.* Winter Navigation Research Board, Report No 43, Helsinki, Finland, 52 pp.

Lau, M. W., Wang, J., and Lee, C. J., 2007. Review of ice modeling methodology. In *Proceedings of the International Conference on Port and Ocean Engineering Under Arctic Conditions.*

Leira, B., Børsheim, L., Espeland, Ø., and Amdahl, J., 2009. Ice-load estimation for a ship hull based on continuous response monitoring, *Proceedings of the Institution of Mechanical Engineers, Part M: Journal of Engineering for the Maritime Environment*, 223(4), 529–540.

Marchenko, N., 2012. *Russian Arctic Seas.* Springer-Verlag Berlin Heidelberg. 10.1007/978-3-642-22125-5

Matala, R., 2023. *Verification of vessel resistance in old brash ice channels through model scale tests*. Aalto University publication series DOCTORAL THESES, 205/2023.

Myland, D., and Hisette, Q., 2021. Experimental investigations on the level ice resistance of ships with non-typical icebreaking bow shapes. *Ships and Offshore Structures*, 16 (sup1), 225–236.

Ralph, F., Ritch, R., Daley, C., and Browne, R., 2003. Use of finite element methods to determine iceberg impact pressure based on internal strain gauge measurements. In *17th International Conference on Port and Ocean Engineering under Arctic Conditions, POAC*, Trondheim, Norway, June 16–19, 2003.

Riska, K., Kujala, P., and Vuorio, J., 1983. Ice load and pressure measurements on board I.B. Sisu. In *7th International Conference on Port and Ocean Engineering under Arctic Conditions*, Espoo, Finland, Vol. 2, 1983, pp. 1055–1069.

Riska, K., Rantala, H., and Joensuu, A., 1990. *Full scale observations on ship-ice contact results from tests series onboard IB Sampo, winter 1989*. Helsinki University of Technology, Laboratory of Naval Architecture and Marine Engineering, Report M-97, Espoo, Finland; 1990.

Ritch, R., 2008. First year hull-ice interaction loads measured on the Louis S. St-Laurent during the 1995 Gulf of St. Lawrence trials. In *The 8th International Conference and Exhibition on Performance of Ships and Structures in Ice, ICETECH 08*, Calgary, Alberta, Canada, July 20–23, 2008.

Schwarz, J., 1983. Sea-ice investigations in the weddell sea during the German Antarctic expedition 1979–80. *Annals of Glaciology*, 4, 246–252.

St. John, J., Daley, C., and Blount, H., 1984. *Ice loads and ship response to ice*. Ship Structure Committee, Report No. 329.

St. John, J., and Minnick, P., 1995. *Ice load impact study on the national science foundation's research vessel Nathaniel B*. Palmer, Ship Structure Committee, Report No. 376.

St. John, J., Sheinberg, R., Ritch, R., and Minnick, P., 1994. Ice impact load measurements abroad the Oden during the International Arctic Ocean Expedition 1991. In *SNAME, 5th International Conference on Ships and Marine Structures in Cold Regions*, Calgary, Alberta, Canada, March 16–18, 1994.

Suominen, M., 2018. *Uncertainty and variation in measured ice-induced loads on a ship hull. Aalto University publication series*. Doctoral dissertation. Aalto University, Helsinki, Finland.

Suominen, M., 2024. *Study on friction coefficient correction for model-scale resistance testing in brash ice channel*, Winter Navigation Research Board, Report No. 130.

Suominen, M., Karhunen, J., Bekker, A., Kujala, P., Elo, M., von Bock und Polach, R., Enlund, H., and Saarinen, S., 2013. Full scale measurements on board PSRV S.A. Agulhas II in the Baltic Sea. In *22nd Interenational Conference on Port and Ocean Engineering under Arctic Conditions*, Espoo, Finland, June 9–13, 2013.

Suominen, M., and Kujala, P, 2012. Ice model tests in compressive ice. In Z. Li and P. Lu (eds.), *Ice Research for a Sustainable Environment: 21st IAHR International Symposium on Ice*, Dalian, China, June 2012.

Suominen, M., and Kujala, P, 2013. A study on measured line load lengths and maximum ice loads on model ship hull. In *Proceedings of the 22nd International Conference on Port and Ocean Engineering under Arctic Conditions, POAC*, Espoo, Finland, June 2013.

Suominen, M., Kujala, P., Romanoff, J., and Remes, H., 2017. The effect of the extension of the instrumentation on the measured ice-induced load on a ship hull. *Ocean Engineering*, 144, 327–339.

Valkonen, J., 2006. *Determination of ship ice load from hull ice damages*. Diplomityö, Teknillinen korkeakoulu, Konetekniikan osasto.

Varsta, P., 1976. *Measurement and Analysis of Ice-Induced Stresses in the Shell of an Icebreaker*. Winter Navigation Research Board, Report No 21, Helsinki, Finland, 1976.

von Bock und Polach, F., Ettema, R., Gralher, S., Kellner, L., and Stender, M., 2019. The non-linear behavior of aqueous model ice in downward flexure. *Cold Regions Science and Technology*, 165, 102775.

Vuorio, J., Riska, K., and Varsta, P., 1979. *Long term measurements of ice pressure and ice-induced stresses on the icebreaker Sisu in winter 1978*. Winter Navigation Research Board, Report No 28, Helsinki, Finland, 50 pp.

8 Numerical Methods

8.1 LOCAL ICE LOADS AND STRUCTURAL RESPONSE

For local load estimation, the primary points of interest are the ice pressure magnitude and distribution on the local structure, which are used to evaluate the capacity of structure to bear the loads for safety concern. Theoretical methods for the estimation of ice loads are established assuming rigid ship hull, simplified ice geometry and idealized ship-ice interaction process, so that the derivation of solutions could be possible. These assumptions are not necessary with numerical modelling. The Finite Element Method (FEM) is the most adopted method, although still challenging, for the estimation of local ice loads on ship structures, due to its rigorous formulation and excellent capability to resolve the stress-strain field and deformation including the cracking process. FEM is especially advantageous in being able to effectively account for the behaviour of ice under tri-axial stress state.

There are essentially two ways to calculate the structural response, here referred to as the fully coupled approach and the weakly- or non-coupled approach. The former attempts to model the full ship-ice interaction process involving both structural deformation and ice failure; the latter simplifies the ice pressure using idealized load patch, e.g., following ice-class rules so no ice failure is modelled. Here we focus on the fully coupled approach since FEA for structural response is a relatively mature topic. The key issues related to FEA of ice-structure interaction are introduced below.

8.1.1 General Setups

8.1.1.1 Modelling of Ship Hull

When the aim is to estimate local ice loads on ship structures, it is typical to model ship structure in a local range (e.g., a stiffened panel between web frames), instead of the whole ship. The ship model can be rigid if the deformation of the structure is expected to be in the elastic range, or elastoplastic if the load level is rather high. The structural members such as the stiffeners and frames shall be modelled if the focus is on the structural response.

8.1.1.2 Ice Geometry

As the local structure of the ship is modelled, correspondingly, a local part of the entire ice sheet is often modelled in the case of colliding a large floe. The contact region can be modelled as a wedge with large opening angle. We recommend the wedge opening angle to be 120°–150° to be consistent with the glancing impact scenario defined by the IACS polar-class rules.

8.1.1.3 Treatment of Water

In the case of ship-ice interaction, the water beneath the ice restricts the deflection of the ice sheet, and hence must be considered. The support from water is often

DOI: 10.1201/9781003676829-8

simplified as a Winkler-type (elastic) foundation, with only hydrostatic force considered. However, this approach cannot account for the effect of bow wave resulting from the ship motion, which may result in underestimation of the ice-load amplitude. In addition, the increase of the water pressure due to fast loading is not included. One approach to account for the hydrodynamic effect is to use the Arbitrary Lagrangian Eulerian (ALE) method.

8.1.1.4 Effect of Dynamics

As mentioned before, the effect of dynamics plays a major role in the amplitude of ice loads. Unless the interaction speed is very low, it is recommended to make dynamic simulation with either explicit or implicit solvers.

8.1.2 Ice Material Model

Defining the proper ice material model is the key to simulate the contact force correctly, and is also the most challenging part of FE analysis of ship-ice interaction. The material model is composed of two parts, namely the constitutive law and the failure criterion. The former defines the stress-strain relationship, while the latter defines the points of yielding and the initiation of cracks. Ice is a viscoelastic material, so we recommend to model the viscosity in the constitutive law. The viscosity, however, is often neglected in the existing work and the deformation of ice is assumed to be rate-independent. The effect of viscosity to ice loads still demands further quantification.

The failure criterion of ice has not reached a consensus in the field of ice mechanics. Ice behaves as a brittle material under fast loading, which is dominated by cracks. However, the most common way to model the failure of ice currently is to use elastoplastic model. The Derradji-Aouat (2003) model has been widely used. This model simplifies ice as a homogeneous, isotropic material. Under complex loading, ice is assumed to yield when the following criterion is reached:

$$\left(\frac{\tau - \eta}{\tau_{\max}}\right)^2 + \left(\frac{p - \lambda}{p_c}\right)^2 = 1 \tag{8.1}$$

where η, $\tau_{\max}$, λ and p_c are constants, τ is the octahedral stress and p the hydrostatic pressure under complex stress state. It can be proved that this model is equivalent to the Tsai-Wu failure criterion for isotropic material when η equals zero. After yielding, ice is allowed to deform plastically, until a failure strain, ε_f, is reached. In addition, for tensile loading, a cut-off stress, σ_c, is defined. Ice is assumed to fail if the tensile stress exceeds this value.

Figure 8.1 demonstrates a FEM simulation domain of a ship bow colliding an ice sheet. As can be seen, the above approach adopts a plasticity methodology, which does not actually capture the crushing process where crushed ice formed and ejected from the contact zone. Nonetheless, this approach provides an engineering way to model the crushing force without resolving the development microcracks. The readers are referred to Liu et al. (2011), Xu et al. (2019) for further reading on this topic. We recommend the readers to be aware of the simplifications and uncertainties

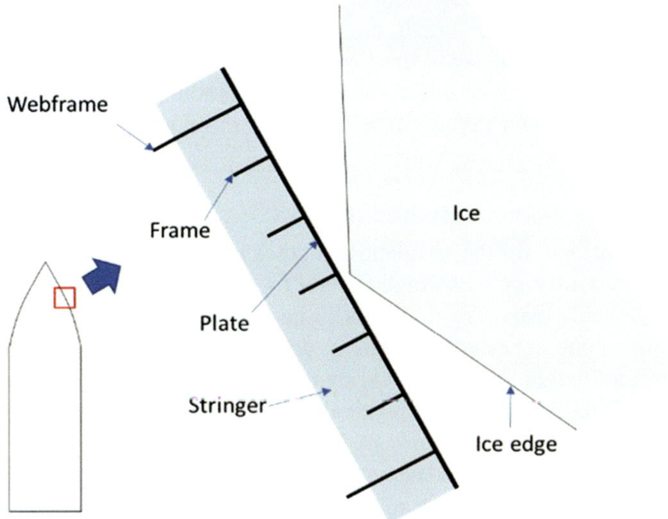

FIGURE 8.1 Illustration of the simulation domain.

when using this ice material model. The constants in Eq. (4.51) need to be carefully selected by calibrating to existing measurements.

It should be noted that, although deployed widely, Eq. (4.51) was proposed for freshwater ice and iceberg ice. With sheet ice, an issue of property variation along the thickness direction arises. Sea ice in nature is typically transverse isotropic, with its strain modulus and strength varied along the thickness due to the gradient of temperature and salinity. The Tsai-Wu failure criterion can handle the transverse isotropy, which is defined as:

$$F_2(\sigma_1+\sigma_2)+F_3\sigma_3+F_{22}\left(\sigma_1^2+\sigma_2^2\right)+F_{33}\sigma_3^2+F_{44}\left(\sigma_4^2+\sigma_5^2\right)+F_{66}\sigma_6^2 \\ +2F_{12}\sigma_1\sigma_2+2F_{23}(\sigma_1+\sigma_2)\sigma_3\leq 1 \tag{8.2}$$

where the F_i s are the material constants; σ_1 and σ_2 are the normal stresses in the horizontal plane and σ_3 the normal stress in the vertical plane; σ_4 and σ_5 are the shear stresses in the vertical plane and σ_6 the shear stress in the horizontal plane (Varsta, 1983). We recommend to account for the transverse isotropy and variation along the thickness when sheet ice is in consideration.

8.1.3 Crack Evolution

There are mainly three methods compatible with FEM which can simulate the propagation of cracks in ice, namely the element deletion technique, the Cohesive Element Method (CEM) and the Extended Finite Element Method (XFEM).

8.1.3.1 Element Deletion Technique

The element deletion (or corrosion) method is based on the simple idea that when the failure criterion is met in an element, the element shall be deleted from the computation field. This can be easily set in most commercial FEM software. The drawback of this method is the violation of mass conservation. In addition, the damage is usually defined by a failure strain (as just introduced) or the damage mechanics, which is incompatible with fracture mechanics.

8.1.3.2 Cohesive Element Method (CEM)

CEM is a powerful tool for the simulation of crack evolution. The method is based on fracture mechanics, making it especially useful for brittle materials such as ice. A layer of cohesive elements with zero thickness is inserted in between the finite elements. The behaviour of the cohesive elements is defined by the Traction-Separation Law (TSL). As exemplified in Figure 8.2, TSL defines the relationship between the traction stress and the separation distance, which can be seen as the opening size of the crack. An initial stress is set in the beginning of the curve, which is fictitious just to avoid infinite stiffness. The crack initiates when the traction reaches the critical level, denoted by σ_f. After that, the crack further opens and the traction it can hold changes accordingly. When the separation reaches the critical value, denoted by δ_m, the cohesive element can bear no more traction and the crack completely opens. The area below the TSL equals the critical energy release rate of ice, denoted by G_c. In this way, the crack can be simulated following the theory of fracture mechanics. As can be noticed, with CEM, the crack can only propagate along the edge of elements. Careful setup of the mesh is therefore important to get correct crack path.

8.1.3.3 Extended Finite Element Method (XFEM)

The Extended Finite Element Method (XFEM), as indicated by its name, is an extension of FEM, which allows discontinuity across crack surfaces by introducing

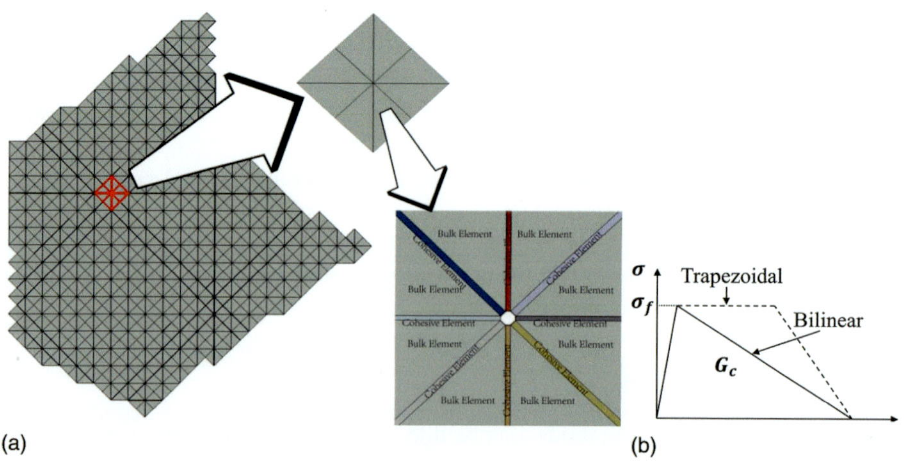

FIGURE 8.2 Example of cohesive elements and the TSL.

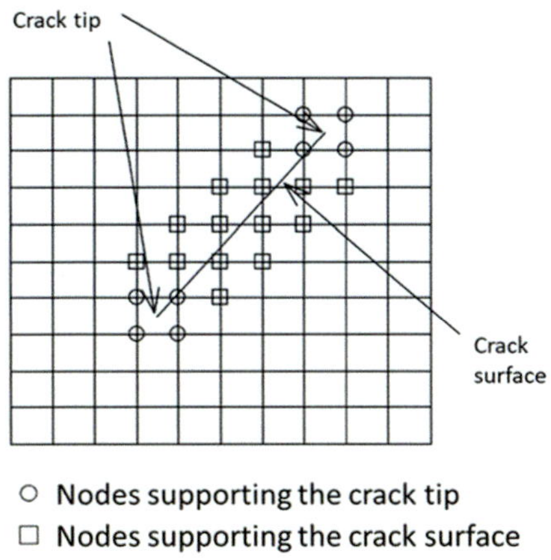

FIGURE 8.3 Illustration of crack propagation across elements with XFEM.

an additional term to the shape function of the elements (Figure 8.3). As a result, cracks are allowed to propagate through elements, instead of going along the edges as with CEM. The property of the cracks can be defined by a traction-separation law, similarly as that by CEM. XFEM allows to study crack growth along an arbitrary, solution-dependent path without needing to remesh the model.

Despite the benefits being mentioned, at the current stage, the maturity of XFEM is not yet high enough to simulate complex crack propagation. Problems such as how to handle multiple cracks encounter still need to be solved. Crack propagation simulation in 3D is not very stable in some commercial software with XFEM. Nonetheless, XFEM is a promising approach with advantageous features, which may be more applicable to ice engineering problems with future development.

In addition to the FEM-based methods, other numerical simulation methods, such as the Discrete Element Method and the peridynamics method, have also been used in ice engineering problems. However, when structural deformation is concerned in local ice loads-related problems, FEM is still the most reliable approach. Methods such as DEM are mostly used in global ice loads-related problems, e.g., estimating ship resistance in ice.

8.1.4 Method of Verification

Before applying numerical methods to simulate ship-ice interaction, it is important to verify that the adopted models can reasonably reflect the engineering properties of ice. This is hereby referred to as verification. The following principles and procedures are recommended to carry out verification.

First, it is recommended to identify what is the critical ice-load limiting mechanism during the ship-ice interaction process. The limiting mechanism can be the crushing failure in case of ship collision with an iceberg, or the flexural failure in case of ship breaking level ice. For the former case, it is recommended to carry out ice indentation tests (e.g., a rigid wall crushing an ice cone or ice ball) and compare with experimental results. For the latter case, it is recommended to carry out bending tests such as three-point or cantilever beam bending, and compare with experimental results.

Besides, it is required to identify what is the type of ice under consideration. Model ice, freshwater ice and sea ice have rather different properties. If the ship-ice interaction problem of concern is regarding sea ice, the verification study must be carried out with sea ice experiments conducted at the same sea region.

To verify the model, one needs to demonstrate that under the simulation setup, ice can exhibit similar macroscopic engineering properties such as the crushing strength, compressive strength, flexural strength, as well as the contact pressure. Afterwards, the simulation setup can be regarded as applicable for the simulation of ship-ice interaction.

8.1.5 Method of Validation

Validation means to confirm the ice-load result as predicted aligns with the experiment. Dedicated experiments shall be carried out for this purpose. Unlike models for ship performance in ice, it is not recommended to use model-scale tests with scaled ice for validation, because artificial ice behave differently from natural ice, which results in completely different local pressure distribution and loading process on the contact area.

Using full-scale measurement results is recommended, but extensive scatter exists in the data, making it challenging to compare directly. Statistical analysis must be carried out to find the correspondence between the measured data and the simulation setup in order to make credible validation. The landing-craft bow test by Varsta (1983) provides useful results for this purpose. In addition, ship-borne ice-load measurement data can be used for validation, but the prevailing ice condition is usually uncertain.

Another way of validation is through lab tests of ice-structure interaction. Examples of such tests are the pendulum tests at Arctec Inc. in Ottawa (Glen & Comfort, 1983), ice crushing tests at NRC in Ottawa (Tuhkuri, 1995), laboratory indentation tests at WARC in Helsinki (Daley, 1990), etc. The benefit is that experimental conditions are well under control. However, these tests put more focus on the ice failure process often using rigid indenter as the structure. It is recommended to carry out fully coupled ice-structure interaction lab test for more credible validation.

8.1.6 Example

Overall, not much work has been carried out using fully coupled simulation. Most of the work focuses on either structure deformation under idealized ice pressure or ice failure process under rigid wall indentation. Liu et al. (2011) carried out FEA of ship collision with an iceberg. Ice is modelled as an elastoplastic material based on Derradji-Aouat (2000) model. Structure members of a ship are modelled and allowed to deform and fracture. The problem is then a shared-energy collision, which means that the kinetic energy is dissipated by both the ship and the ice. Figure 8.4 shows the collision process, presenting details of the structure deformation and ice breaking.

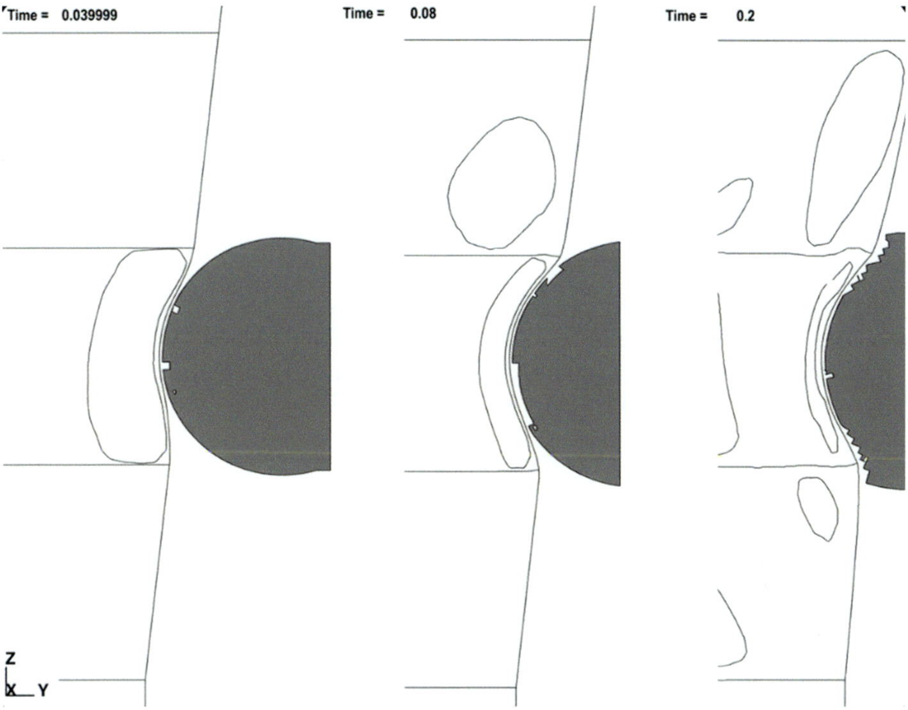

FIGURE 8.4 Simulation of ship collision with an iceberg by Liu et al. (2011).

8.2 SHIP PERFORMANCE

Unlike simulations of local ship-ice interaction, to simulate a ship going through an ice field, the scale increases from metres to hundreds of metres. The scenario to be simulated then involves a large number of local ship-ice interactions, which significantly complicates the problem. Moreover, the process also involves the interaction between discrete ice bodies, including broken ice pieces, ice floes, ice rubbles, etc. FEM-based methods typically cannot handle these complexities effectively and efficiently. Methods developed for the purpose of ship performance evaluation can be divided into two categories, here referred to as the empirical-analytical numerical method and the computational solid mechanics method.

8.2.1 Empirical-Analytical Numerical Method

Methods in this category model local ship-ice interaction process analytically, usually combined with empirical factors, instead of computationally. Li (2020) made a comprehensive analysis on the composition and methodology of such methods. The benefit is the significantly reduced computational demand as the algorithm does not need to resolve the stress-strain field. The accuracy of ship performance evaluation then depends majorly on the analytical characterization of the ship-ice interaction process.

Su (2011) developed a numerical simulation approach, which has been developed by multiple researchers (Tan 2014; Zhou et al., 2016) since then, and named as the Circumferential Crack Method (CCM) by Zhou et al. (2020). Su (2011) uses semi-empirical formulae to model the icebreaking process. The bearing capacity of ice edge, denoted by P_f, is calculated as:

$$P_f = C_f \left(\frac{\theta}{\pi}\right)^2 \sigma_f h_i^2 \tag{8.3}$$

where θ is the wedge opening angle; σ_f is the flexural strength of ice; h_i is the thickness of ice; C_f is an empirical coefficient. When the contact force exceeds this value, an ice piece will break from the intact ice sheet. The size of the ice piece is characterized by an arc with radius R being

$$R = C_l l_c (1 + C_v v_n) \tag{8.4}$$

where l_c is the characteristic length of the ice sheet as introduced in Chapter 3; v_n is the normal contact velocity; C_l and C_v are empirical coefficients. With these two formulae, local icebreaking process can be calculated in the simulation model. The established numerical model can simulate ship performance in straight ahead/astern and turning operation in level ice, providing evaluation of resistance, attainable speed and maneuverability (see Figure 8.5 for an illustration).

Another way of modelling the icebreaking process is to derive analytical formulae. Lubbad and Løset (2011) developed the Simulator for Arctic Marine Structures (SAMS) using Nevel's wedge bending solution. The basic theory of Nevel's solution has been introduced systematically in Chapter 4, with which one can calculate the maximum stress in the ice field analytically. Li et al. (2018) extended Nevel's wedge beam to wedge plate, which increases the fidelity of icebreaking simulation. SAMS has been developed further in recent years to include radial-crack algorithm based on fracture mechanics. Figure 8.6 shows the mechanism involved in SAMS simulation by Lubbad et al. (2015).

The third way of modelling the icebreaking process is through the so-called meta-modelling approach. The core idea is to see icebreaking simulation as a cross-scale problem, involving ship scale, local scale and material scale. High-precision computations (e.g., FEA) are performed on local ship-ice interaction in a parametric way,

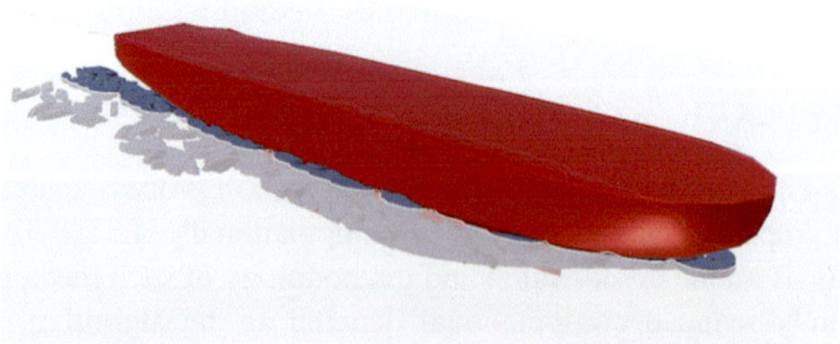

FIGURE 8.5 Numerical simulation of ice breaking straight ahead using CCM.

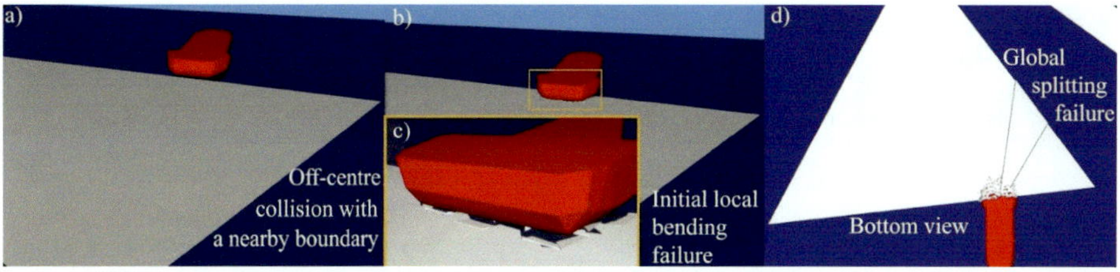

FIGURE 8.6 Mechanisms involved in SAMS (Lubbad et al., 2015).

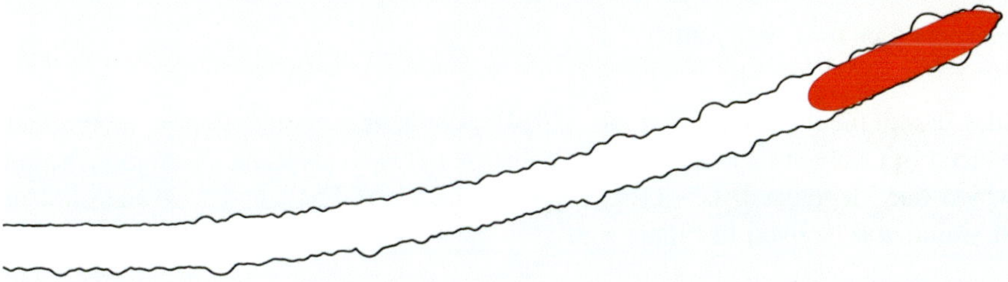

FIGURE 8.7 Simulation of a ship turning in ice by ASOIS.

so that analytically formed equations can be established on the resulting simulation dataset. The equations describe the mapping between local-scale parameters such as ice geometry as well as contact force and material-scale responses such as the distribution of stress, acting similarly as closed-form solutions. A numerical simulation model can then be established based on the equations. The benefit of metamodelling approach is to maintain high accuracy while minimizing computation. Li et al. (2019) developed this concept using parametric FEA and neural network, and established the Aalto Ship-Operation-in-Ice Simulator (ASOIS, Li et al., 2020). Figure 8.7 presents a snapshot of ship turning simulation using ASOIS.

In addition to the modelling of ice breaking, it is of at least equal importance to model the collision between ship and ice as well as between discrete ice pieces. A widely adopted method to model collisions between ship and ice objects is the impulse-based method, often referred to as the Non-smooth Discrete Element Method (NDEM). Unlike the Discrete Element Method (DEM) which will be introduced soon, NDEM calculate the collision between objects using impulse. It does not directly resolve the contact force, which avoids the complexity of contact modelling, therefore computes fast. The drawback is then that local contact force needs to be calculated from impulse by estimating the collision time, which makes it inaccurate although equivalent from energy perspective. Examples of models using this approach include the SAMS model mentioned above and the model by Yang et al. (2021). Besides using

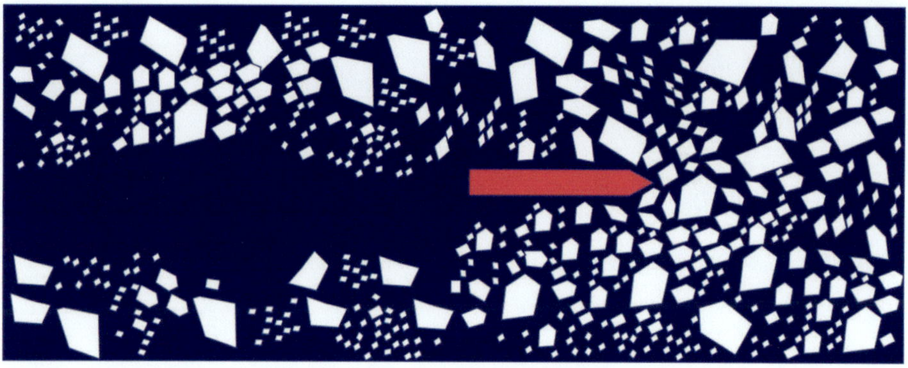

FIGURE 8.8 A snapshot of GEM simulation of a ship in broken ice field.

(Re-produced from Daley et al., 2014.)

impulse-based method, Daley et al. (2014) adopts the classic Popov method as introduced in Chapter 4 to calculate the collision between ship and ice objects, based on which they developed their GPU-event-mechanics (GEM) model. A snapshot of GEM simulation is given in Figure 8.8.

8.2.2 Computational Solid Mechanics Method

Methods in this category computes the entire stress-strain field and resolve ice deformation and failure process at the material scale using computational solid mechanics (CSM) methods. The most adopted method in the field of ice engineering is the Discrete Element Method (DEM). As indicated by its name, DEM is especially capable and initially developed for the contact mechanics between discrete particles. Nonetheless, in recent years, DEM has also been applied to model ice breaking using the concept of particle bonding.

The main difference between DEM and NDEM is that it resolves the contact between bodies and calculate the contact force based on the contact geometry and penetration, while NDEM is based on impulse. Figure 8.9 presents the concept. The two particles overlap with a penetration δ, based on which the normal and tangential contact force, F_n and F_t, can be calculated. To form an ice piece with arbitrary geometry, one just needs to pile up the particles and connect them with bonds, as illustrated in Figure 8.9. The bond can be set to be breakable, making it possible to simulate process of ice breaking. Beside round particles, other shapes such as squares, round disks and polyhedral can also be used as the discrete particles.

The normal contact force between discrete elements is often solved using an elastic–viscous or elastic–viscous–plastic model. The tangential force is usually calculated by Coulomb friction model. When ice breaking needs to be simulated, one needs to set the strength of the connecting bonds and calibrated with mechanical tests. Here we do not aim to make a thorough introduction of the basic theory of DEM, but refer the reader to existing review articles on this topic, such as Tuhkuri and Polojärvi (2018). Due to the need of small contact area detection, DEM simulation typically requires small-time steps and can be computationally very expensive.

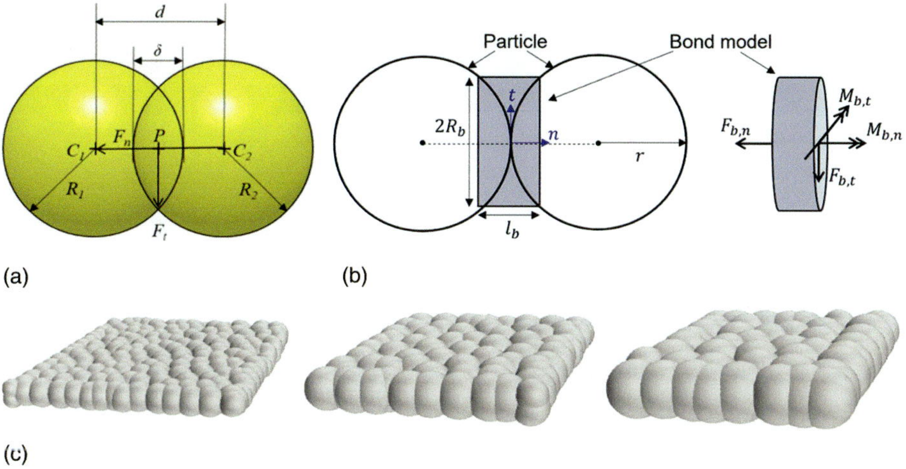

FIGURE 8.9 Concept of DEM. (a) Contact between two particles; (b) bonded particles; and (c) piling up of particles to form ice floes of different thicknesses (Huang et al., 2022).

DEM outperforms other CSM-based methods (such as FEM) in the simulation of ship performance in broken ice fields. This includes at least ice floes, brash ice, ice ridge rubble, as well as broken ice pieces during continuous ice breaking. Li and Huang (2022) presented a systematic review of DEM application in the evaluation of ship performance in broken ice fields. To date, most of the published work on DEM application in ice is conducted with commercial software, such as Star-CCM+ and LS-Dyna. In spite of that, some universities have developed their in-house DEM code. For example, the DEM code by Aalto University has been initially developed for structure-ice interaction, and later been used for ships in ridged ice field and floe ice field (see Figure 8.10, Gong, 2021). The ICEDEM model by Dalian University of Technology has been developed for the simulation of ice breaking by ships and offshore structures, and has been benchmarked with experiments and full-scale ship tests (Ji et al., 2020) (Figure 8.11).

Fluid dynamics plays an important role in the motion of broken ice pieces. This has been increasingly investigated during recent years, mainly by coupling of

FIGURE 8.10 Simulation of a ship going through an ice ridge by Aalto in-house DEM (Gong, 2021).

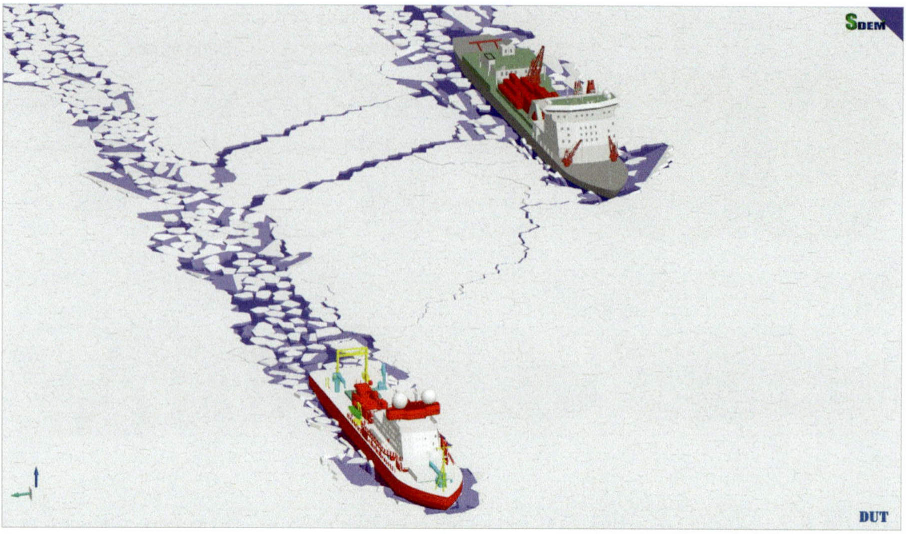

FIGURE 8.11 Simulation of two ships sailing in parallel by ICEDEM (Ji et al., 2020).

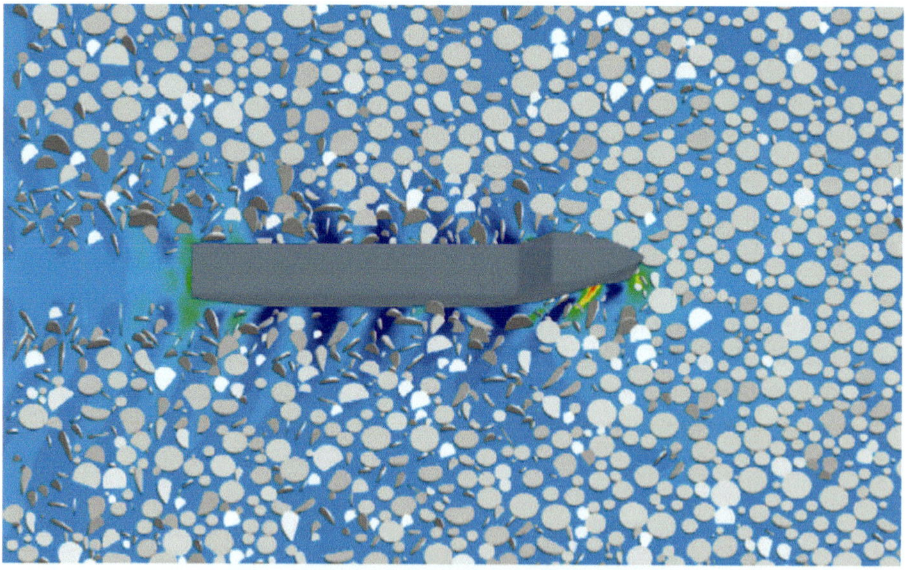

FIGURE 8.12 Coupled CFD-DEM simulation of a ship sailing through ice floes (Huang et al., 2020).

Computational Fluid Dynamic (CFD) and DEM. Many people use commercial software to establish CFD-DEM simulations of ships in broken ice. An example of such simulations is given in Figure 8.12. The effect of ship-generated waves on the motion of ice pieces can be observed.

In addition to DEM, other CSMs such as FEM and peridynamics have also been applied to ship performance evaluation in ice. These methods, however, have not yet reached the maturity of DEM on handling both ice breaking and discrete contacts.

8.3 PROPELLER ICE LOADS AND PROPULSION PERFORMANCE

Similar to simulations of ship hull interaction with ice, simulations of propeller-ice interaction can also be divided into two categories based on the purpose. The first category aims to evaluate strength of propeller blades for safety consideration. Extreme scenarios, typically the ice milling scenario, are considered with large and thick ice blocks. The target is to get the maximum loads on the blades and the stress at the blade roots. The second category aims to evaluate propulsion performance, especially the change of propulsion efficiency, due to the presence of ice blocks. The target is to evaluate the cavitation performance and extra power required compared with propulsion in open water.

Modelling of ice deformation and failure in the simulation of propeller-ice interaction is similar to that in ship-ice interaction. Cohesive element method (CEM) is often adopted for this purpose with FEM, e.g., by Zhou et al. (2019). In addition, peridynamics has been applied by Ye et al. (2017), which shows promising feature. Figure 8.13 shows some simulations of ice milling by propeller blades with CEM and peridynamics. The phenomena of ice cutting can be captured by both methods. Nonetheless, it should be noted that existing ice material models are usually developed at low contact speeds, while when propeller cuts ice, the speed can be much higher than ship speed. Thus, ice modelling is the major uncertainty and source of deviation associated with such simulations.

When ice blocks are in the vicinity of the propeller, it blocks the inflow and changes the flow field around the propeller, which affects its cavitation performance.

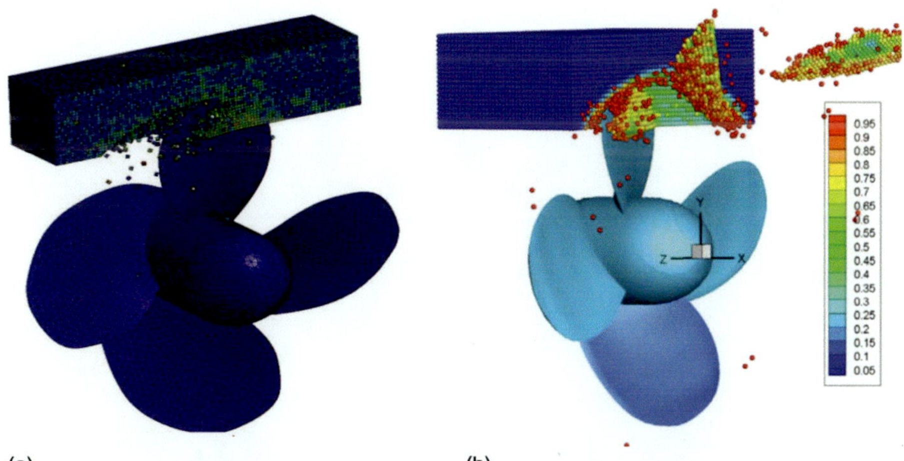

FIGURE 8.13 Simulation of propeller milling ice by CEM (left, Wang et al., 2019).

This is often termed as *ice blockage*. Simulation of such scenarios needs to be carried out with CFD. Figure 8.14 shows the experimental and CFD simulation by Zheng et al. (2024) on this problem. It was found that the effect of ice blockage is two-fold. On one hand, it reduces the inflow velocity of propeller, which increases its propulsion performance. On the other hand, it induces larger area of cavitation, which reduces propulsion performance.

To evaluate overall propulsion performance of a propeller behind a ship, coupled CFD-DEM simulation is a viable approach. This is then similar to coupled CFD-DEM simulations of ship sailing through ice, and is also conducted with commercial software. The overset meshing technique is typically needed to model the rotating propeller. Figure 8.15 shows the simulation by Xie et al. (2023) about a self-propelled ship through brash ice. In such simulations, ice bodies are commonly assumed as non-breakable rigid bodies. This can introduce considerable deviation and uncertainty in the results.

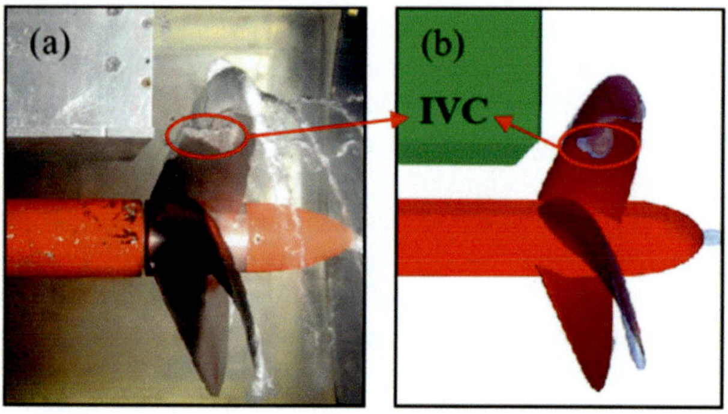

FIGURE 8.14 Ice blockage and cavitation on the blade in (left) an experiment and (right) corresponding simulation (Zheng et al., 2024).

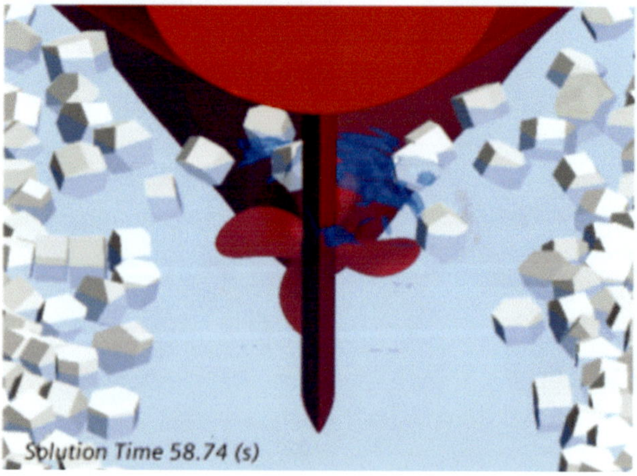

FIGURE 8.15 Coupled CFD-DEM simulation of self-propelled ship in brash ice, showing details of propeller-ice collision (Xie et al., 2023).

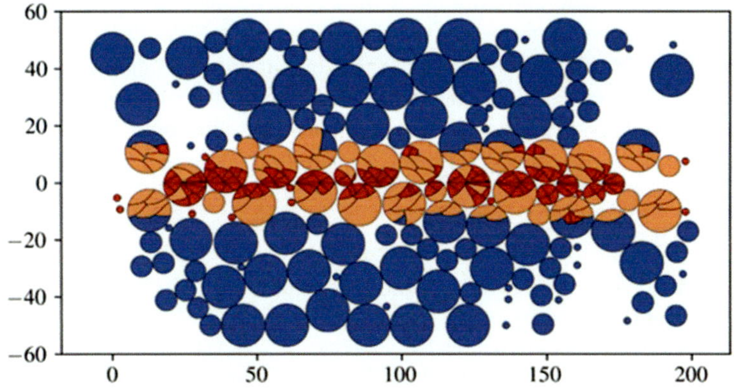

FIGURE 8.16 Breaking pattern of floe icebreaking simulation. Measures on both axes are in m. Red floes are assumed to interact with the propeller, orange floes are pushed aside the ship and blue floes are not in contact with the ship's hull (Lund et al., 2025).

In-house simulation codes have been developed by some researchers for propeller-ice interaction-related problems. Wang et al. (2006) presented their PROPICE model, which can simulate ice loads on propeller blades as well as propulsion performance. The advantage of this work over commercial software is that both hydrodynamic effects and ice failure process are considered. Semi-empirical methods were incorporated into the model to estimate the ice-crushing pressure. The simulation results compare well with experimental results. Here we recommend the readers to refer to Wang et al. (2006) for further knowledge on this topic.

As discussed already in Section 3.4, the validation of the numerical simulation with full-scale data is extremely important. In a recent paper by Lund et al. (2025), an existing failure model for ice based on the Mohr–Coulomb nodal split approach is modified, based on measurements conducted in the North Pole region, to better represent the lower strength of polar ice compared to laboratory ice. The modified material model for the sea ice is used to compute the propeller torque and a load spectrum for the propeller-ice interaction of the research ship *S.A. Agulhas II* and the results are compared with full-scale torque data with good correlation. To this end, a floe-icebreaking simulation is used to estimate the size and shape of the ice cusps hitting the propeller of this ship and a set of finite element simulations of the propeller-ice interaction utilizing the modified Mohr–Coulomb nodal split model is conducted, see Figure 8.16.

8.4 SUMMARY

This chapter offers a general introduction of numerical simulation methods related to polar ship technology. Overall, numerical simulation methods have great potential in handling structure-ice and propeller-ice interaction problems, due to their capability in material modelling and hydrodynamic computation. However, the complexity of ship-ice-water coupling makes it very difficult to establish credible numerical simulation models. One should start with a clear understanding of the modelling purpose, based on which necessary simplifications can be identified. For example, it has been shown through above examples that evaluations of safety and performance require rather different simulation setups. One should bear in mind that numerical simulation models should not be as complex and comprehensive as possible. As Einstein said: things should be made as simple as possible, but not simpler.

REFERENCES

Daley, C. G., 1990. Ice edge contact – A brittle failure process model. Doctor Thesis. Helsinki University of Technology.

Daley, C., Alawneh, S., Peters, D., and Colbourne, B., 2014. GPU-event-mechanics evaluation of ice impact load statistics. In *OTC Arctic Technology Conference 2014 Feb 10*, pp. OTC-24645. OTC.

Derradji-Aouat, A., 2000. A unified failure envelope for isotropic fresh water ice and iceberg ice. *Proceedings of the ETCE/OMAE-2000 Joint Conference, Energy for the New Millennium*, New Orleans, USA, OMAE2000/P&A-1002.

Derradji-Aouat, A., 2003. Multi-surface failure criterion for saline ice in the brittle regime. *Cold Regions Science and Technology*, 36(1–3), 47–70.

Glen, I., and Comfort, G., 1983. Ice impact pressure and load: investigation by laboratory experiments and ship trials. In *Proceeding 7th International Conference on Port and Ocean Engineering under Arctic Conditions (POAC)*, Helsinki, vol. I, pp. 516–533

Gong, H., 2021. Discrete-element modelling of ship interaction with unconsolidated ice ridges: Ridge resistance and failure behaviour. Doctor Thesis. Aalto University.

Huang, L., Li, F., Li, M., Khojasteh, D., Luo, Z., and Kujala, P., 2022. An investigation on the speed dependence of ice resistance using an advanced CFD+ DEM approach based on pre-sawn ice tests. *Ocean Engineering*. 264, 112530.

Huang, L., Tuhkuri, J., Igrec, B., Li, M., Stagonas, D., Toffoli, A., Cardiff, P., and Thomas, G., 2020. Ship resistance when operating in floating ice floes: A combined CFD & DEM approach. *Marine Structures* 74, 102817.

Ji, S., Liu, L., Ji, S., and Liu, L., 2020. DEM analysis of ice loads on offshore structures and ship hull. *Computational Granular Mechanics and Its Engineering Applications*, 2020, 237–310.

Li, F., and Huang, L., 2022. A review of computational simulation methods for a ship advancing in broken ice. *Journal of Marine Science and Engineering*, 10(2), 165.

Li, F., Kõrgesaar, M., Kujala, P., and Goerlandt, F., 2020. Finite element based meta-modeling of ship-ice interaction at shoulder and midship areas for ship performance simulation. *Marine Structures*, 71, 102736.

Li, F, Kotilainen, M, Goerlandt, F, & Kujala, P., 2018. A new icebreaking pattern for the application in numerical simulation of ship performance in level ice. In *Proceedings of the ASME 2018 37th International Conference on Ocean, Offshore and Arctic Engineering. Volume 8: Polar and Arctic Sciences and Technology; Petroleum Technology*, Madrid, Spain. June 17–22, 2018. V008T07A017.

Li, F., Kotilainen, M., Goerlandt, F., and Kujala, P., 2019. An extended ice failure model to improve the fidelity of icebreaking pattern in numerical simulation of ship performance in level ice. *Ocean Engineering*, 176, 169–183.

Liu, Z., Amdahl, J., and Løset, S., 2011. Integrated numerical analysis of an iceberg collision with a foreship structure. *Marine Structures*, 24(4), 377–935.

Lubbad, R., and Løset, S., 2011 A numerical model for real-time simulation of ship-ice interaction. *Cold Regions Science and Technology*, 65(2), 111–127.

Lubbad, R., Løset, S., Lu, W., Tsarau, A., and van den Berg, M., 2015. An overview of the Oden Arctic Technology Research Cruise 2015 (OATRC2015) and numerical simulations performed with SAMS driven by data collected during the cruise. *Cold Regions Science and Technology*, 156, 1–22.

Lund, J., Sapp, L., Kubiczek, J., Böhm, A. M., and von Bock und Polach, F., 2025. Simulated ice loads on a ship propeller and comparison with full-scale measurements. *Marine Structures*, 104, 103879.

Su, B., 2011. Numerical predictions of global and local ice loads on ships. Doctor Thesis. Norwegian University of Science and Technology.

Tan, X., 2014. Numerical investigation of ship's continuous-mode icebreaking in level ice. Doctor Thesis. Norwegian University of Science and Technology.

Tuhkuri, J., 1995. Experimental observations of the brittle failure process of ice and ice-structure contact. *Cold Regions Science and Technology*, 23, 265–278.

Tuhkuri, J., and Polojärvi, A., 2018. A review of discrete element simulation of ice-structure interaction. *Philosophical Transactions of the Royal Society A: Mathematical, Physical and Engineering Sciences*, 376(2129), 20170335.

Varsta, P., 1983. On the mechanics of ice load on ships in level ice in Baltic Sea. Doctoral dissertation, VTT Technical Research Centre of Finland.

Wang, J., Akinturk, A., and Bose, N., 2006. Numerical prediction of model podded propeller-ice interaction loads. In *Proceedings of the 25th International Conference on Offshore Mechanics and Arctic Engineering. Volume 2: Ocean Engineering and Polar and Arctic Sciences and Technology*, Hamburg, Germany. June 4–9, 2006. pp. 667–674.

Wang, F., Zou, Z. J., Li, Z., Wang, Y., Yu, H., and Zhang, H., 2019. Numerical simulation of ice milling loads on propeller blade with cohesive element method. *Brodogradnja: An International Journal of Naval Architecture and Ocean Engineering for Research and Development*, 70(1), 109–128.

Xie, C., Zhou, L., Ding, S., Liu, R., and Zheng, S., 2023. Experimental and numerical investigation on self-propulsion performance of polar merchant ship in brash ice channel. *Ocean Engineering*, 269, 113424.

Xu, Y., Hu, Z., Ringsberg, J. W., and Chen, G., 2019. Nonlinear viscoelastic-plastic material modelling for the behaviour of ice in ice-structure interactions. *Ocean Engineering*, 173, 284–297.

Yang, B., Sun, Z., Zhang, G., Wang, Q., Zong, Z., and Li, Z., 2021. Numerical estimation of ship resistance in broken ice and investigation on the effect of floe geometry. *Marine Structures*, 75, 102867.

Ye, L. Y., Wang, C., Chang, X., and Zhang, H. Y., 2017. Propeller-ice contact modeling with peridynamics. *Ocean Engineering*, 139, 54–64.

Zheng, S., Huang, Q., Zhou, L., and Li, H., 2024. Numerical investigation on high-fidelity flow field of an ice-class propeller in ice blockage and cavitation flow. *Ocean Engineering*, 301, 117626.

Zhou, L., Diao, F., Song, M., Han, Y., and Ding, S., 2020. Calculation methods of icebreaking capability for a double-acting polar ship. *Journal of Marine Science and Engineering*, 8(3), 179.

Zhou, Q., Peng, H., and Qiu, W., 2016. Numerical investigations of ship–ice interaction and maneuvering performance in level ice. *Cold Regions Science and Technology*, 122, 36–49.

Zhou, L., Wang, F., Diao, F., et al., 2019. Simulation of ice-propeller collision with cohesive element method. *Journal of Marine Science and Engineering*, 7(10), 349.

Appendix I:

Summary of icebreakers in service and under construction

APPENDIX I

Summary of icebreakers in service and under construction

Vessel Name	Nationality	L (m)	B (m)	T (m)	Displacement (t)	Deadweight (t)	Propulsion Power (kW)	Propulsion Thrusters	Ice Class	Max Continuous Icebreaking (m)
In Service										
Yamal	Russia	148	30	11	23,460	2,750	54,000	Three-shaft nuclear-turbine-electric	RMRS Arctica	3
Viktor Chernomyrdin	Russia	146.8	29	9.7	23,152	8,321	25,000	Diesel-electric	RMRS Icebreaker8	3.5
Alexander Sannikov	Russia	121.7	25	8	13,397	4,581	21,500	Wartsila	RMRS Icebreaker8	2
Aker Arc 130A	Russia	121.7	25	8	Unknown	Unknown	21,500	Four main diesel generators and one harbour generator	RMRS Icebreaker8	2
Moskva	Russia	114	27.5	8.5	14,300	7,243	16,400	Diesel-electric	RMRS Icebreaker6	1.5
St. Petersburg	Russia	114	27.5	8.5	14,300	7,107	16,400	Diesel-electric	RMRS Icebreaker6	1.5
Vaygach	Russia	151.8	29.2	9	21,000	3,550	36,800	Three-shaft nuclear-turbine-electric	RMRS LL2	2
Taymyr	Russia	151.8	29.2	9	21,100	3,550	36,800	Three-shaft nuclear-turbine-electric	RMRS LL2	2.2
Arktika	Russia	173.3	34	10.5	33,530	9,000	60,000	Three shafts, single rudder	RMRS Icebreaker9	2.8
Sibir	Russia	173.3	34	10.5	33,540	9,000	60,000	Three-shaft nuclear-turbine-electric	RMRS Icebreaker9	2.8

(*Continued*)

Appendix I (Continued)

Vessel Name	Nationality	L (m)	B (m)	T (m)	Displacement (t)	Deadweight (t)	Propulsion Power (kW)	Propulsion Thrusters	Ice Class	Max Continuous Icebreaking (m)
Ural	Russia	173.3	34	10.5	33,540	9,000	60,000	Three-shaft nuclear-turbine-electric	RMRS Icebreaker9	2.8
Project 10570	Russia	152	31	8.5	20,700	Unknown	40,000	Unknown	LK-40 class	2.4
Gennadiy Nevelskoy	Russia	104	21	7.9	3,824	3,670	15,600	Diesel-electric drive	RMRS Icebreaker6	1.5
Novorossisk	Russia	119.8	27.5	8.5	14,334	5,142	17,400	Diesel-electric	RMRS Icebreaker6	1.5
Murmansk	Russia	119.8	27.5	8.5	14,334	5,048	17,400	Diesel-electric	RMRS Icebreaker6	1.5
Vladivostok	Russia	119.8	27.5	8.5	14,334	5,610	17,400	Diesel-electric	RMRS Icebreaker6	1.5
Sevmorput	Russia	260.3	32.2	11.8	61,880	33,980 (summer); 26,480 (Arctic)	29,420	Single-shaft steam turbine	RMRS UL	1.8
Alexey Chirikov	Russia	99.83	21.2	7.9	10,708	4,191	13,000	Diesel-electric drive	RMRS Icebreaker6	1.7
Vitus Bering	Russia	99.9	22.1	7.9	10,708	4,158	13,000	Diesel-electric drive	RMRS Icebreaker6	1.7
Akademik Tryoshnikov	Russia	133.53	23.25	8.5	16,539	6,634	14,000	Twin-shaft diesel-electric	RS Arc7	1.1
Varandei	Russia	100	21.7	9.5	7,338	3,150	Unknown	Diesel- electric	RS Arc7	1.5
Vladislav Strizhov	Russia	99.3	19	7.6	5,871	3,867	20,000	Diesel- electric	RS Arc7	Unknown

Yuri Topchev	Russia	99	19	7.6	5,871	3,949	20,000	Diesel- electric	RS Arc7	Unknown
Pacific Endeavor	Russia	91,5	19	7.6	4,992	3,700	14,000	Diesel- electric	RMRS Icebreaker6	1.5
Pacific Endurance	Russia	91.5	19	7.6	4,992	3,700	14,000	Diesel- electric	RMRS Icebreaker6	1.5
Kapitan Sorokin	Russia	141	31	7.8	15,385	5,138	63,000	Three-shaft diesel-electric	KM * LL3	1.5
Akademik Fedorov	Russia	141.2	23.5	8.5	16,200	7,200	Unknown	Diesel-driven	RS ARC7	2
Kapitan Dranitsyn	Russia	129.02	26.54	8.5	14,917	4,515	16,200	Three-shaft diesel-electric	RMRS LL3	1.6
Kapitan Khlebnikov	Russia	129.2	26.5	8.5	14,917	4,515	16,200	Three-shaft diesel-electric	RMRS LL3	1.6
Kapitan Nikolayev	Russia	121.15	26.69	8.5	16,017	4,836	16,200	Three-shaft diesel-electric	RMRS LL3	1.6
Yermak	Russia	97.5	21.6	7.3	8,730	Unknown	Unknown	Four-shaft	Unknown	Unknown
Krasin	Russia	134.84	26.05	11	20,247	7,554	26,500	3 x Stromberg DC motors	LL2	1.8
Admiral Makarov	Russia	134.84	26.05	11	20,247	7,560	26,460	3 x Stromberg DC motors	LL2	1.8
Fedor Ushakov	Russia	99.9	21.2	7.9	12,450	3,670	15,600	Diesel-electric drive	icebreaker6	1.7
Stepan Makarov	Russia	99.9	21.2	7.9	12,450	3,670	15,600	Diesel-electric drive	icebreaker6	1.7
Mikhail Lazarev	Russia	99.9	21.2	7.9	12,450	3,670	15,600	Diesel-electric drive	icebreaker6	1.7
Nikolay Zubov	Russia	114	18	6	6,800	Unknown	12,600	Diesel-electric	RS Arc7	1.7
Aleut	Russia	86.7	19.5	7.1	7,441	2,600	4,400	Diesel-electric	DNV ICE-10	0.8
Pomor	Russia	86.7	19.5	7.1	7,466	2,600	4,400	Diesel-electric	DNV ICE-10	0.8
Norman	Russia	86.7	19.5	7.1	7,441	2,600	4,400	Diesel-electric	DNV ICE-10	0.8

(Continued)

Appendix I (Continued)

Vessel Name	Nationality	L (m)	B (m)	T (m)	Displacement (t)	Deadweight (t)	Propulsion Power (kW)	Propulsion Thrusters	Ice Class	Max Continuous Icebreaking (m)
Baltika	Russia	72.64	20.5	7	5,708.8	1,150	7,500	Triple podded oblique acting icebreaker	RMRS Icebreaker6	1
SCF Sakhalin	Russia	99.9	20.95	7.5	9,980	4,298	15,200	Diesel-electric	DNV ICE-10 Icebreaker	1.7
Ivan Papanin	Russia	114	18	6	6,800	Unknown	12,600	Diesel-electric	RS Arc7	1.7
Vasiliy Golovnin	Russia	164	22	6.7	10,800	10,700	Unknown	Unknown	Unknown	Unknown
Dikson	Russia	92	21.2	6.8	6,583	2,281	7,648	Twin-shaft diesel-electric	LL4	1
Mudyug	Russia	92	21.2	6.5	8,154	2,920	7,648	Twin-shaft diesel-electric	LL4	1
Magadan	Russia	92	21.2	6.5	6,210	1,909	7,648	Twin-shaft diesel-electric	LL4	1
Ob	Russia	89	20	7.5	7,613	2,567	12,000	Diesel- electric	RS Icebreaker	1.5
Dudinka	Russia	87	21	6	Unknown	1,219	Unknown	Diesel- electric	Unknown	Unknown
Tor	Russia	85	21	6.5	Unknown	1,024	Unknown	Diesel- electric	Unknown	Unknown
Vladimir										
Ignatyuk	Russia	88	17.82	8.3	7,077	2,076	17,063	Twin-shaft controllable-pitch propellers	RMRS icebreaker7	1.2
Polar Sea	United States	122	25.45	9.4	10,863	7,000	56,000	Three-shaft diesel-electric/gas combination	Polar class icebreaker, USCG	1.8

Polar Star	United States	122	25.45	9.4	10,863	7,000	56,000	Three-shaft diesel-electric/gas combination	Polar classi icebreaker, USCG	1.8
Healy	United States	128	25	8.92	16,257	Unknown	22,000	Twin-shaft diesel	Medium icebreaker, USCG	1.4
Nathaniel B. Palmer	United States	94	18.3	6.8	6,909	2,500	13,000	Diesel- electric	ABS A1	1
Aiviq	United States	110	24.4	8.6	12,892	4,129	21,000	Diesel- electric	ABS A3	1.5
Harry DeWolf class	Canada	103.6	19	5.7	6,615	Unknown	9,000	Twin-shaft diesel-electric	PC5	1
Margaret Brooke	Canada	103.6	19	5.7	6,615	Unknown	9,000	Twin-shaft diesel-electric	PC5	1
Max Bernays	Canada	103.6	19	5.7	6,615	Unknown	9,000	Twin-shaft diesel-electric	PC5	1
CCGS Terry Fox	Canada	88	17.82	8.3	4,234	2,113	Unknown	Twin-shaft	CASPPR Arctic Class 4	Unknown
Louis St. Laurent	Canada	119.8	24.38	9.91	11,345	4,640	20,142	Three-shaft diesel-electric	PC 4	Unknown
CCGS Vincent Massey	Canada	83.7	18	7	6,872	2,600	13,440	Twin-shaft	DNV ICE 10	1
CCGS Captain Molly Kool	Canada	83.7	18	7	6,872	2,600	13,440	Twin-shaft	DNV ICE 10	1
CCGS Jean Goodwill	Canada	83.7	18	7	6,872	2600	13,440	Twin-shaft	DNV ICE 10	1
Polaris	Finland	110	24	8	10,961	2 × 400	19,000	Triple podded	PC 4	1.8
Nordica	Finland	116	26	8	9,392	Unknown	15,000	Twin-shaft diesel-electric	Polar 10	0.8
Fennica	Finland	116	26	8	9,392	Unknown	18,450	Twin-shaft diesel-electric	PC3	1.8
Kontio	Finland	99	24.2	8	9,130	Unknown	15,000	Twin-shaft diesel-electric	PC4	0.8

(Continued)

Appendix I (Continued)

Vessel Name	Nationality	L (m)	B (m)	T (m)	Displacement (t)	Deadweight (t)	Propulsion Power (kW)	Propulsion Thrusters	Ice Class	Max Continuous Icebreaking (m)
Otso	Finland	99	24.2	8	Unknown	Unknown	15,000	Twin-shaft diesel-electric	PC 4	0.8
Oden	Sweden	107.8	31	7	9,438	Unknown	5,400	Twin-shaft diesel-electric	Polar 20	1.9
Frej	Sweden	104.6	23.8	7.3	9,500	2200m3	16,200	Twin-shaft diesel-electric	Polar 20	Unknown
Atle	Sweden	104.6	23.8	7.3	9,500	2200m3	16,200	Twin-shaft diesel-electric	Polar 20	Unknown
Sisu	Finland	105	24	7.7	9,660	Unknown	16,200	Twin-shaft diesel-electric	Polar 20	Unknown
Urho	Finland	105	24	7.7	9,660	Unknown	16,200	Twin-shaft diesel-electric	Polar 20	Unknown
Ymer	Sweden	104.7	23.8	7.3	9,500	2,259m3	Unknown	Unknown	Polar 20	1.2
Njord Viking	Norway	85	22.8	6.6	6,279	Unknown	Unknown	Unknown	Ice-1A	Unknown
Loke Viking	Norway	85	22	6.6	6,372	Unknown	Unknown	Unknown	Ice-1A	Unknown
Brage Viking	Norway	85	22	6.6	Unknown	Unknown	Unknown	Unknown	Ice-1A	Unknown
Magne Viking	Norway	85	22	6.5	Unknown	Unknown	Unknown	Unknown	Ice-1A	Unknown
KV Svalbard	Norway	103.7	19.1	6.5	6,375	Unknown	10,000	Twin-shaft diesel-electric	Polar 10	Unknown
Kronprins Haakon	Norway	100.38	21	8.66	9,000	Unknown	13,200	Twin-shaft diesel-electric	PC3	1
Xue Long 2	China	122.5	22.3	7.9	14,300	4,500	15,000	Twin-shaft diesel-electric	PC3	1.5

Xue Long	China	166.4	22.6	9	21,025	8,759	9,100	Single-shaft	CCS B1	1.1
Haibing 722	China	103.1	18.4	5.16	7,000	4,860	Unknown	Unknown	TC6	1
Haibing 723	China	103.1	18.4	5.16	7,000	4,860	Unknown	Unknown	TC6	1
RV Tan Suo San Hao	China	104	19.7	6.7	9,300	3,100	12,000	Diesel- electric, 2 Pod	PC4	1.2
Shirase (AGB-5002)	Japan	134	28	9.7	11,600	Unknown	22,064	Diesel-electric	Unknown	1.5
Shirase (AGB-5003)	Japan	138	28	9.2	20,000	Unknown	22,000	Twin-shaft diesel-electric	Unknown	1.5
Agulhas II	South Africa	134.2	21.7	7.65	13,687	4,780	12,000	Twin-shaft diesel-electric	PC5	1
RSV Aurora Australis	Australia	94.91	20.3	7.86	8,158	3,911	Unknown	Diesel-electric	1 A Super	1.23
RSV Nuyina	Australia	160	25.6	9.3	24,000	9,005	Unknown	CODLAD system	PC3	1.65
ARA Almirante Irízar	Argentina	121.3	25	9.2	14,899	4,600	11,900	Twin-shaft diesel-electric	Unknown	1
RV Araon	South Korea	111	19	6.8	6,950	2,901	10,000	Twin-shaft diesel-electric	Polar 10	1
Sir David Attenborough	United Kingdom	128.9	24	7	12,790	4,475	5,500	Twin-shaft diesel-electric	PC4	1
Polarstern	Germany	117.9	25	10.7	17,300	4,374	15,000	Unknown	Diesel electric	1.5
Under Construction										
Lider-Class Icebreaker	Russia	209	47.7	13	71,400	100,000	120,000	Four shafts, twin rudders	Unknown	4.3
Polar Security Cutter	United States	140	27	10.5	23,300	Unknown	Unknown	Twin pods + shaft propellers	PC2	2.5
John G. Diefenbaker	Canada	158	28	10.5	23,700	Unknown	34,000	Single Pod + twin shaft propellers	PC2	2.5
CCGS *Arpatuuq*	Canada	138.5	29.4	10	22,800	Unknown	30,000	Diesel- electric, 2 Pod, one shaft line	LR, PC2	2.0

(Continued)

Appendix I (Continued)

Vessel Name	Nationality	L (m)	B (m)	T (m)	Displacement (t)	Deadweight (t)	Propulsion Power (kW)	Propulsion Thrusters	Ice Class	Max Continuous Icebreaking (m)
CCG Multipurpose Icebreaker (MPI)	Canada	99.9	20.3	6.4	9,000	Unknown	7,200	Diesel- electric, azimuth	PC4	1.0
RV Polarstern II	Germany	160	27.3		Unknown	Unknown	28,500	Diesel- electric, azimuth	PC2	Unknown
RV Araon 2	South Korea	111	19	9.9	7,507	Unknown	15,000	Diesel- electric, 2 Pod	PC3	1.5
Arctic Security Cutter	Unites States	99.9	21	7.0	9,000	Unknown	13,000	Diesel- electric, azimuth	PC3	1.5
Arctic Security Cutter	United States	100	20.4	6.4	9,000	Unknown	7,200	Diesel- electric	LR, PC4	1.0
New icebreaker	Finland	96	24	7-8	7,500	Unknown	10,500	Diesel- electric	PC5+	Unknown

In addition, Sweden has ordered an A-class icebreaker from South Korea.

Index

Pages in *italics* refer to figures and pages in **bold** refer to tables.

I

J

K

L

M

N

P

For Product Safety Concerns and Information please contact our EU representative GPSR@taylorandfrancis.com Taylor & Francis Verlag GmbH, Kaufingerstraße 24, 80331 München, Germany

Batch number: 10777146

Printed by Printforce, the Netherlands